THE CHANGING FACE OF CALL

LANGUAGE LEARNING AND LANGUAGE TECHNOLOGY

Series Editors:
Carol A. Chapelle, *Iowa State University, Ames IA, USA*
Graham Davies, *Thames Valley University, London, UK*

THE CHANGING FACE OF CALL: A JAPANESE PERSPECTIVE

EDITED BY

Paul Lewis
Aichi Shukutoku University, Nagoya, Japan

SWETS & ZEITLINGER PUBLISHERS

LISSE   ABINGDON   EXTON (PA)   TOKYO

Library of Congress Cataloging-in-Publication Data

Applied for

Printed in the Netherlands by

Copyright © 2002 Swets & Zeitlinger B.V., Lisse, The Netherlands

Published by: Swets & Zeitlinger Publishers
www.szp.swets.nl

ISBN 90 265 1934 6 (HB)
    90 265 1935 4 (PB)
ISSN 1568-248X

Contents

3   The CALL Classroom and Beyond

Preface

This collection of papers began after the JALT CALL SIG[1] conference held at Kanto Gakuen University, Gunma, in May, 2001. The authors of most of the papers in this book participated in the conference, entitled "The Changing Face of CALL: Emerging Technologies, Emerging Pedagogies", and it was at that time that the idea of compiling a volume of papers was proposed. We are very proud to bring you this collection of ideas and research presented at the conference. Hopefully it will shed some light on the present direction of CALL in Japan, complementing the first book in this series, subtitled "A European Perspective".

[1] Japan Association for Language Teaching CALL Special Interest Group

Acknowledgements

I would like to express my appreciation to the large team who cooperated in putting together this book. Firstly, I would like to thank everyone at Swets and Zeitlinger Publishers, Marc Weide, Léon Bijnsdorp, Arnold Veenhoff, and Patrick Gordijn for their tireless and patient support and professional guidance throughout.

I am also indebted to the JALT CALL SIG team, especially my Associate Editor, Stephen Shucart, for his unfailing assistance, and to all the reviewers and proofreaders: Charles Adamson, Paul Allum, Frank Berberich, Malcolm Field, Jane Hoelker, Lawrie Hunter, Douglas Jarrell, David Kluge, Malcolm Kruse, Rick Lavin, Jan Minagawa, Barry Natusch, Linh Pallos, and all the authors who assisted in so many ways. I am very grateful also to the coordinator of the SIG, Tim Gutierrez, for giving the go-ahead for the project, and to Odin Dekkers for his invaluable help in arranging publication at the outset.

Finally, special thanks to Hiroko and Alex for putting up with yet another CALL book; without their help and patience, this book would truly not have been possible.

Nagoya, June 2002

Paul Lewis
Editor

Introduction

Paul Lewis
Aichi Shukutoku University

Over the last few decades, Japan has gained a reputation as something of a trailblazer in the development of technology. However, Japan also remains a country somewhat off the beaten tourist track, expensive to visit, and farther from Europe and the US than mere geographical distance would suggest. Somehow, Japan's image as a land of mystery extends to developments in CALL, especially since much of the educational technology in use in this country is not exported, and also because a substantial portion of the literature on the implementation of CALL is only written in Japanese.

While the computer itself did not originate in Japan, it has certainly found a home here. Writing about Japan's history of educational borrowing, Bryn Holmes notes, "It is illuminating to consider whether the computer is merely another example of an introduction of new technology . . . or, instead whether it has followed a new route and imposed its own culture" (1998, p. 182). This present volume is intended to help the reader consider this question by shedding light on theoretical research, practical resources, and the changing role of the CALL classroom in Japan, all from the perspective of Japan's CALL practitioners, both native and non-native speakers of English.

The book is divided into three sections: "Theories, models, and paradigms", "CALL resources", and "The CALL classroom and beyond". To a large extent, any division of a book such as this – especially one concerning such a relatively young field of study – will necessarily involve overlap. To some extent, every paper falls into all three categories, but in the process of drawing up the book, these sections seemed to present themselves quite naturally, and we felt that they represent three distinct directions currently being taken by CALL in Japan.

Section One concerns "Theories, models, and paradigms". The first paper, by Malcolm Field, introduces two new theoretical models for use with tertiary level EFL studies, derived from a web-based bulletin board research. As with every paper in this book, while the initial focus is on the Japanese context, there is likely to be much crossover, and many parallels drawn with situations in other parts of the globe. We hope these papers will stimulate comparative and cooperative follow-up studies by other researchers.

Kevin Ryan then takes us through a brief history of hypertext theory, ending at the point in time where the WWW begins. He then proceeds to integrate hypertext concepts into classroom practice, reporting on some powerful tools for classroom use. Next, Frank Berberich looks at artificial intelligence and the rapidly developing area of natural language processing; Yoshihiko and Shizuko Ariizumi also take up this aspect of adaptivity in a paper which examines CALL from the viewpoint of the Japanese as a Second Language educator. Lawrence Dryden and Michelle Morrone's paper maintains that thread, linking the increasingly popular multiple intelligence theory to CALL, and finding that the theory offers explanatory power and solutions for some of the specific problems faced by Japanese language learners.

The final paper in this section, by Charles Adamson and Stephen Shucart, attempts to find the CALL equivalent of what is known in physics as the "theory of everything". Adamson and Shucart argue that, especially in Japan, much language education is not theory-driven, due to the absence of a coherent theory of the process and nature of that learning. Working with concepts developed in complexity theory, they outline what such a theory may look like and then apply it to CALL. The key concepts of emergence, learning, acquisition, and fitness are brought into play in theory that must be grounded at the DNA level yet be able to guide a computer programmer to create maximally efficient language learning software. They sketch a long future path for the application of this theory, and they urge that work begin as soon as possible on filling in the details and applying the theory.

Section Two, "CALL resources", focuses on some of the computer-based tools now becoming available in Japan and worldwide. Introducing this section, Monika Szirmai surveys corpus linguistics and its associated tools in the Japanese context. Although it appears that corpus studies already have a long tradition in Japan, application to CALL is underdeveloped at present, and Szirmai calls for corpus linguistics to be introduced into teacher training programs so that upcoming generations of teachers may include it in their theoretical and practical repertoire.

The next two papers deal with vocabulary. Bradley Saunders presents a database program used with students of technical writing, noting particularly the autonomy it gives to the learners. John Paul Loucky, on the other hand, examines the various types of computerised bilingual dictionaries available, comparing traditional book dictionaries with hardware, software, and Web-based versions, claiming that the electronic versions could have a profound influence on future CALL.

Peter Wanner introduces a new use of the CHILDES software, known for its application to the transcription of spoken dialogue. In Wanner's study, however, the students themselves did the transcription, with the aim of increasing spoken fluency. While the results tentatively suggest some benefit, Wanner calls for a more detailed analysis in the interests of confirmation and illumination. Interestingly, and perhaps counter to the perceived diligence of the stereotypical Japanese student, Wanner also notes some abuses of the transcription process, and describes some measures he was forced to take to prevent this. Wanner's study is not unusual in this debunking aspect; throughout this book, in many ways, stereotypes about Japan are shown to be invalid or outdated.

The final paper in this section, by James Duggan, looks at security resources available for CALL teachers and administrators, and provides many URLs for further reference and the acquisition of helpful utilities. While not directly connected with the specific act of teaching, it is clear that the CALL teacher has a responsibility to maintain the integrity of the CALL system and the privacy of students' work.

It is perhaps the face of the classroom that is showing the greatest change. Thus, the chapters in Section Three, "The CALL classroom, and beyond", seek to redefine not only what is done and what should be done within the four walls of the classroom, but also to deconstruct those walls in search of virtual and global learning environments.

Douglas Jarrell gives us a brief history of Japan's very tardy adoption of computers in education, a fact that may surprise many. He proceeds to discuss many of the aspects of network-based language teaching, which are taken up and developed in other chapters in this section as well. Jarrell's conclusion is that computer-mediated communication offers a particularly well-suited solution to the problems specific to the Japanese context. The next two papers, both by Japanese CALL researchers, examine cultural aspects of language learning. Kazunori Nozawa focuses on his keypal exchanges, and Kenji Kitao explores the use of student webpage projects. Nozawa finds mixed results from his initial experiments with keypal exchanges, and gives the reader a number of reasons why this is so. Kitao provides numerous URLs for real examples of student work, so that the interested reader can examine not just the product of Kitao's classes, but also that of other innovative classes around the world.

Michael Kruse continues with an in-depth analysis of the new computer-based TOEFL test, recently adopted in Japan to replace the paper-and-pencil version. He examines possible motivations behind the use of computer-adaptive testing, and offers some speculation on TOEFL's underlying algorithms. As Jarrell points out, computer literacy is often surprisingly low in Japan, and Kruse examines the effect of this factor on the Japanese TOEFL candidate, and its impact with the new, mandatory writing paper. At the time of writing, the future of the computer-based TOEFL in Japan is far from certain; this paper offers a good perspective on reasons for this uncertainty.

The final two papers take CALL out of the classroom in a physical sense, but also move tentatively beyond CALL itself. Patricia Thornton and Chris

Houser investigate mobile education, the use of handheld devices to enhance learning. While the annoyance and distraction of the *keitai denwa* (mobile phone) in class is familiar to anyone teaching in Japan nowadays, these authors, and David Kluge in the final chapter, offer a rationale for turning this into a new form of learning technology. While Thornton and Houser highlight the difference between push and pull media, Kluge goes further, and proposes HALL – handheld-assisted language learning, which would overcome many of the physical limitations inherent of current "place locked" CALL classrooms.

In presenting these seventeen papers, we strongly support Angela Chamber's (2001) call in the first book in this series for "controlled empirical studies and the need for greater use, accessibility and flexibility in the technology which is available." We sincerely hope that the present volume supports the progress of wise and effective technology implementation in the classroom. If wisdom prevails, a bright future awaits.

References

Chambers, A. (2001). Introduction. In A. Chambers & G. Davies (Eds.), *ICT and language learning: A European perspective.* Lisse: Swets and Zeitlinger.

Holmes, B. (1998). From black ships to microchips: The arrival of ICT in Japan. In P. N. D. Lewis (Ed.), *Teachers, learners, and computers: Exploring relationships in CALL* (pp. 181-188). Nagoya: Chubu Nihon Kyouiku Bunkakai.

Section 1:

Theories, Models, and Paradigms

1

Towards a CALL Pedagogy: Student Use and Understanding

Malcolm H. Field
University of Cambridge

1. Introduction

The use of information and communication technologies (ICT) in language education has been influencing the nature and pedagogy of language teaching in Japan for more than a decade. Nelson (1998), for example, states that "Rapid changes in communication and information technologies are revolutionizing education and providing new tools to customise learning environments" (p. 90). The appearance of Web based instruction and Web based teaching (WBI/WBT) is providing opportunities to transform the process and pedagogy of foreign language delivery.

In recent years, the explosion of interest in ICT in foreign language teaching and education has not been unanimously accepted. Optimists (Sakamoto, 1992; Van Dusen, 1997; Bennett, 1999) tend to consider ICT as the saviour for teaching, whereas realists (Knobel, Lankshear, et al., 1998; Selwyn, 1999) are cautious to proclaim new learning and educational advantages based on much of the current empirical evidence. Bennett (1999), for example, argues that education will change because computers can teach better than most human teachers by the use of proper software, and there is a need throughout the world for improved education. "School will be lively and vibrant. Students will be energized while learning rapidly, well and enjoyably . . . . machines will teach but humans will educate." However, the mere provision of a computer with high speed Internet access does not automatically imply effective instructional activities or quality instruction (Blair, 1996; Bannan-Ritland,

Harvey, & Milheim, 1998). Furthermore, the use of ICT in the classroom is constrained by sociocultural variables, such as the role of the educational institution within the society, the culture of teaching and education at the institution and the beliefs of the student and teacher (Cuban, 1986; Warschauer, 1998).

Researchers argue (Laurillard & Marullo, 1993; Kuramoto, 1999) that ICT provide a means to facilitate the L2 learner's internal thought processes. It also acts as a psychological tool, enabling the individual to engage in the higher mental functioning that comes from interactions with social life *a la* Vygotsky (1978). It is argued that ICT has the potential to become the medium within the Zone of Proximal Development (ZPD) that promotes learning. The ZPD refers to the cognitive development that occurs in a social nature with cognitive support from a more capable person (Jones & Mercer, 1993). That is, language education, and in particular second or foreign language acquisition, is an active process involving interaction with other people and things.

CALL is the merging of ICT and language education. An issue in teaching approaches and pedagogy with CALL is the conflict of providing only computer-mediated instruction and not an environment of collaborative communication. Marjanovic (1999) argued that "rather than adopting old teaching methods along with new information technologies, it is necessary to investigate new previously unknown possibilities offered by new technologies and design new methodologies for learning and teaching" (p. 138).

This paper is an attempt, using student voices, to provide models for CALL in language education in Japan. For this paper, language education refers to English as a foreign language (EFL) education for the Japanese student.

In Japan, ICT as an information and communication resource in education, particularly language education, has seemingly been overlooked. Previous Prime Minister Mori, and Sony Corporation chairman Idei presented a plan to steer Japan into becoming the most advanced ICT nation in the world within five years. The draft sets out four key areas for the government to deal with. They include (a) the building of infrastructure for ultra-high speed connections [undefined]; (b) the creation of an environment which will foster e-commerce; (c) the implementation of the "so-called electronic government"; and (d) the training of ICT engineers and instructors (*Panel unveils 5-yr IT strategy plan*, 2000). The focus was, however, on economics and the financial sector.

A reason for this oversight could be as Bachnik (1999) has explained: "ICT poses such a challenge to Japan [because] the system itself is precisely what is at stake here. The changes required fly directly in the face of organisational continuity . . . . by prioritising skill-levels and individual competencies, over institutional and positional rankings" (p. 13). Similarly, Narita (1999) argues that the "main barrier against educational applications of the Internet . . . is a widespread underestimation of the importance of information technologies among many of the top administrators" (p. 1). Against such obstacles, the provision of ICT based pedagogy in language education, such as CALL, is not always widely supported.

A further issue for the gap in use of ICT in education in Japan, has been the way the computers that have filtered into classroom practices have been used. For example, Sugimoto (1999) points out from his research, that his students felt that the computer classes they took in their freshman year were useless and they had forgotten all they had learned there (a reflection of the traditional rote-learning drill pedagogy). Generally, ICT has been presented under the curriculum of computer-skills, rather than as multimedia or information and communication technologies.

These issues have led to research in Japan being undertaken by teachers who utilise their own classroom and students to generate data. These studies are valuable for teachers involved in CALL education and provide an excellent springboard for other research. Access, control, class size, group dynamics, training, and time will continue to be the main factors influencing the extent to which students use computers (Sloane, 1997; McMahon, Gardner, et al., 1999).

A positive development in CALL research has been a transformation from youthful awe and excitement about new educational tools to a mature and directed pedagogical and philosophical focus with questions raised and answers sought. These questions include: Does the mere provision of a computer mean that students will acquire a new (second or foreign) language? If a new (second or foreign) language is acquired by students because of interactions with a computer, what kind of language will be acquired? What are the theoretical, educational, and language objectives in CALL? And, are these objectives acceptable and understood by everyone in the institution? These questions provide scope for future research. This paper is a step in the process of trying to address some of these questions.

2. Research overview

This paper presents data generated from a doctoral dissertation undertaken at the University of Cambridge. Research investigated the influence of ICT on language learning and use for Japanese EFL university students. The students were in their first year at a mid-sized private university in central Japan. The university provided excellent computing facilities and all students had access to an Apple Macintosh laptop computer and to desktop computers in laboratories. Methods of research included the use of written protocols, questionnaires, email communication, interviews, and bulletin board interaction. Data for this paper was generated from student postings to a web-based bulletin board (BBS). Students were asked to express their attitudes and opinions about using CALL in their EFL studies. Two classes participated in the BBS discussion.

A BBS is a valuable tool for both synchronous and asynchronous communication. It is beneficial for language education as it allows the teacher options of observation or participation, language evaluation, and feedback. The BBS also facilitates a student centred language generation and learning philosophy. One weakness is that it requires keyboard skills and some degree of computer literacy. As this paper later emphasizes, students with low ICT skills may need to enrol in a class specifically designed to improve such skills.

Two distinct approaches were used for the BBS communications. The first was a structured approach, in which students received a fixed question and were asked to respond to it. The students were not encouraged to respond to each other's postings. The second approach was semi-structured. Several general categories were posted and students were encouraged to respond to them and to each other. The researcher also participated in this interaction.

This two-step process proved the most effective in allowing students to (a) develop confidence and competence with the tool, and (b) develop thought patterns and language in the target language. Open flexible use of a BBS provides other rewards and advantages; however, in the initial stages, structured and semi-structured approaches helped students stay focused on the target language and tasks. This discussion is derived from student postings during the structured approach.

3. Demographics

One hundred and ten first year students were involved in the research; however, only two classes – a total of thirty-nine students – participated in the BBS discussion. Seventy-one percent of students in the total research group acknowledged using a computer at home, and 86% had used a computer at the university. Female student response frequency to using the computer at home and university was higher compared to the male students (74% - 66% at home, 88% - 83% at university). Male students had a higher response to using the Internet than female students (66% - 71%). More female students, however, stated that they had written or read email (74% - 59%). This discrepancy between the frequency of Internet use and email use has been accounted for by the ability to log on to an email account without opening an Internet browser.

The higher computer use by female students at the university of study in Japan is contradictory to results of studies conducted in the United Kingdom (Selwyn, 1998). The research data does not yield sufficient information to fully account for this statistic. However, comments generated at an interview with one of the university CALL teachers may help explain the results.

> [The female students] make very good teachers to the slower boys – which tends to be my group that is not as fast – not all boys, of course, as it is mixed – but that tends to be the way our university has pulled in students. I think that the better boys are going to the better universities and there are not as many girls in those better universities. So we get a very high level girl – the type of girl that should be in the better universities but hasn't applied and they haven't applied for social reasons – they are not encouraged to go further – to their full potential. (Private communication)

In other words, the teacher believes that "high level" male students are more attracted to the higher level or status national or private universities, whereas female students, because of different social and cultural expectations, do not feel pressure to enter these educational institutions. The result, she argues, is that her university faculty attracts a higher academic female compared to the male.

4. The data

Interactions with the BBS were conducted in class for nearly two months. The structured approach to using the BBS consisted of various questions being posted to the

BBS and students being asked to respond to them. One question posted by a CALL teacher asked students to evaluate CALL classes as a process in their EFL studies.

A common theme expressed by all BBS postings was the "computer" as opposed to the "English" nature of the CALL class. In other words, students commented either positively or negatively about the use of computers, and the degree to which they believed they had mastered that tool.[1]

> I felt that this cause is how usefull for larning English. And I was thinking that it is wast of time to use computer. But now I've seemed to find the purpouse this <teacher's name> cause. That is this cause's main purpouse is to get used to using computer and Internt. That is why we have another English Communication Skills which is <name>. First, I did not like this cause very much. The reason is I can not use computer fluently. All of student was seemed to be able to use computer not difficulty. But still now, it can't be said that I can use the computer. First, I would like to ask you wethre that (this main purpouse is to get used to use computer and Internet) is correct or not? . . . Last, I want to know what you are going to do at this cause next term?   MN (F)

There was clear evidence of a gulf between student expectations about the CALL class and its curriculum, pedagogy, and objectives. Their comments were not specific to one CALL class teacher. Evidence from the BBS postings highlights that the students were either: (a) confused about the purpose of a CALL class, (b) believed it was a computer skills class, or (c) wanted more of a speaking component to the course.

The majority of female students posted messages emphasizing that the course was both interesting and fun, albeit hard at times. Nevertheless, nearly every female student separated the computer/Internet components of the class from the EFL component.

> I like this course and teacher, because computer and teacher are very fun, this course is difficult at first. But I take Computer Communication Appreciation class. So, this course is easy for me. I want to study more computer. SH (F)

> I didn't like the computer. But, this lesson is good. So, I like the computer. I remembered how to inter-net. It's very exciting! Now everyday I use my computer. Thank you!   YKat (F)

> Through this class, I get many things. First I don't like a computer a lot. It is very difficult and not convenient for me. Morever teacher explained in English. I was very trouble. But I get thinking power. As a result, I think that This class Was good. AIw (F)

> Especially, the Internet was very very difficult!!.. I think I have one English class a week, so I want to take more and more English classes. So, to understand the Enternet home pases is very hard for me. I think computer communication needs skill of English. If I can speak English like <teacher>, I could use a computer more . . . NO (F)

Male students overwhelmingly commented about the difficulty of CALL classes. This revolved around two points: students' ability to use the computer, and their English language proficiency.

> I learned how to use the computer this clsaa. FOr example,link,copy, paste and so on. So I think it got used to do the computer,and it used the computer very well. TT (M)

> I think that this course is too difficult for me to understand English . . . I couldn't understand this class. And I can't use a computer well . . . YN (M)

> I often use a computer to Internet and email. But I don't know how to use a computer still. So I thought useing a computer in a English class is very very difficult. SK (M)

> I think this course is interesting. But Sometime this course is difficulut for me. I did not undersand What I doing this course. It is interesting to use computer. I tink I want to more study English and comruter. Anyway I think very interesting this course! KKar (M)

Despite the difficulties expressed by male students, nearly everyone included a favourable comment about the CALL classes. These suggest that many students enjoyed utilizing ICT within their EFL studies and it is, therefore, an educational tool that still needs to be considered, especially in language acquisition and learning.

5. Analysis

The concern by students about their difficulty experienced in CALL classes due to English proficiency or computing abilities should be important to teachers. It was outlined above that the majority of students already had some experience with computers, the Internet, and email. Those with no experience therefore had no schemata on which to base learning choices or decisions. In other words, CALL classes are all-English classes and students having had no experience with computers or ICT need to negotiate meanings not only in new English words (usually jargon), but also in their understanding of the computer and possibly even the Internet. Simultaneous, and in some cases competing, demands are being made upon students' cognitive skills and repertoire of learning strategies from three distinct, though related, objects of study: (a) the EFL curriculum, (b) the practice environment of the computer, and (c) the use of ICT as an information resource.

What can be concluded from an analysis of the BBS postings is that not all students find a CALL class suitable for foreign language learning. Students may need to be given the option of using ICT as part of their EFL studies or otherwise, as some students may prefer to study in language classes that do not use ICT as a tool of instruction. Students who prefer computer-mediated communication may, therefore, more readily utilise the opportunities that teachers provide through the use of ICT in a CALL class.

Each student must make meta-cognitive choices and prioritise learning choices and strategies. Some students have established schemata and preferred learning strategies for using a computer in their studies. For these students, it is less problematic to engage with the new English (of instruction) in a CALL class. However, most students have never needed to consider alternative or simultaneous learning strategies. In other words, their experience up until the end of secondary education is learning new subject matter in their native Japanese. If students have experienced an all-English class given by a native speaker, they have usually had the safety of translations into or from Japanese. Moreover, on these occasions, the all-English curriculum attempts to repeat and reinforce subject matter already covered by Japanese teachers of English, and build on already established schemata.

Furthermore, Japanese students generally do not approach their EFL learning with the purpose of language production and communication. The primary objective for most students is to attain a satisfactory examination result to graduate from high school, enter university or postgraduate school, or graduate from university. Therefore, many students may find it difficult, apart from the cultural language-use phenomena, to "create" a communicative syntax and structure that can be understood by other interlocutors.

A further obstacle to be recognized is that in any one class, not only do student skill levels with ICT vary, but also English language proficiencies are vastly different. In one classroom, a teacher may have highly able students that have lived abroad for more than one year, together with others who have difficulty going beyond drills and structured patterns. The same holds true for ICT skills; levels can reflect personal home use, previous high school opportunities and pedagogy, or attitude, and skills may vary from low proficiency (unable to use a keyboard or mouse or complete simple word processing tasks) to high, whereby a student can perform simple programming tasks, such as using html.

The BBS comments and the mixed ICT skills and EFL proficiencies highlight the need to separate students according to both English proficiency and ICT experience/skills. If CALL is to have the positive impact on language learning, and provide the advantages that some research has claimed, students will need to be organized according to both proficiencies. Once foreign language and ICT proficiency is adequate, the student may appreciate the advantages of language learning through ICT, particularly in a CALL classroom.

Student comments to the BBS provide data that has led to the development of theoretical models: A theory of curriculum development and a theory of learner development. These two models are discussed below.

6. A theory of curriculum development

The theory of curriculum development (Fig. 1) is developed from the assumption that the student is at an entry level into both the EFL and CALL curriculum. The students may, therefore, be in their first year or in graduate school. It is more probable, however, that they will be first years, commencing compulsory tertiary level EFL education. The discussion for the model is developed from this premise.

Two separate though interrelated courses need to be offered to the students, and

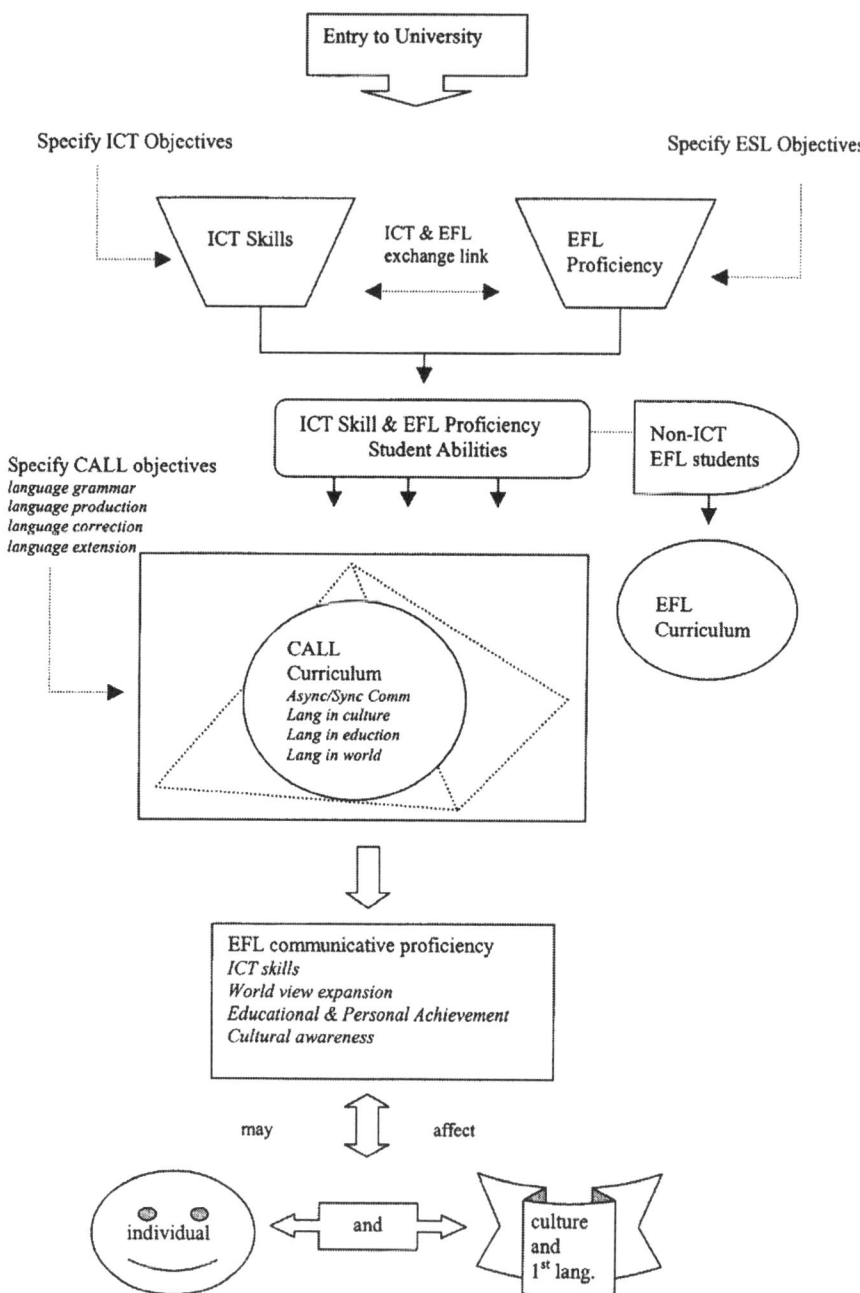

Figure 1. The theory of curriculum development.

their objectives clearly specified. Detailing objectives may enable the students to make informed choices before enrolling, and furthermore, enable them to plan how to engage with the new course material, such as adopting or seeking new learning strategies to process information.

The first of these courses is an ICT skills course, which should be conducted in the first language, Japanese. Students at many universities in Japan, including the one providing access for this study, first experience ICT as an information and communication tool in their EFL CALL classes. Students at the university of the study were able to enrol in an additional computer appreciation class, though few of the research group chose this option. ICT should, therefore, be made compulsory in the academic study program to ensure that student skill level more closely matches that of other members of any one class.

It is anticipated that most students will develop the requisite computer literacy within the next five years (as the policy of computer education becomes a fundamental element of education in senior high, junior high, and also in the primary school systems). Currently, however, many first year students seem to lack basic computer literacy. Providing ICT skills in all-English classes, which brings its own rewards, may also create conflicting and competing demands upon the student's repertoire. Furthermore, as the support (linguistic, administrative, academic, governmental, and cultural) for all-English content-based classes is low, the benefits of providing such classes are negated. To expedite the learning process and adoption of ICT skills, the course should be provided in Japanese.

The second course should be an all-English EFL course with a focus on general EFL proficiency. Students should attain a (university determined) proficiency in the target language before being allowed to enrol in a CALL based EFL curriculum. Similarly, an ICT skills level should be attained before enrolment in the EFL CALL based curriculum. English based ICT skills, such as typing a document in English, or searching for an English language based Internet site, should be taught within the classes, and these activities should be related to the EFL proficiency curriculum.

Once students have attained the requisite EFL proficiency and ICT skills, they should be given the option of continuing their EFL university studies in either classes that utilise ICT (CALL) or in those that do not. Students choosing not to enrol in an EFL CALL class can, therefore, continue EFL studies using traditional language teaching pedagogies and learning tools. Students choosing to enrol in classes with a CALL curriculum could then draw on schemata developed from their previous two distinct, though related, courses.

The objectives of CALL need to be made clear to students. CALL is fundamentally about the development, acquisition, and production of language, and therefore, language grammar, production, correction, and extension need to be the foci of the curriculum, pedagogy, and all activities. Students, furthermore, need to recognise and understand these objectives and may need regular reminders. When students fail to understand or lose sight of these objectives, they may begin to believe that the objective of CALL is the development of ICT skills, as may have been the case for students participating in this study.

Pedagogy and activities within a CALL class should be developed around these aforementioned language objectives. This may include asynchronous and synchronous communication activities, such as email or BBS interactions; or focusing on language in culture, education, and in the world, such as by interacting with websites, communicating through MOOs and chat, or through video conferencing sessions. The curriculum, objectives, and pedagogy should, however, be based on the prism model.

The prism model (Fig. 2), developed as an extension of the Sapir-Whorf hypothesis (Whorf, 1940, 1956), proposes that an individual's understanding of the world, and ability to adjust to that world is influenced by education, language, culture, and thinking. Expanding any one side of the prism enlarges its internal space as all its sides are inextricably linked. This inner space is the individual's world-view, ability to adjust and adapt to the world, and process information about that world. This is often called adaptability, flexibility, and growth. Growth is case and individual specific. The secondary objectives of a CALL curriculum are to increase the individual's world-view through the acquisition of language, exposure to new cultures, thinking, and education.

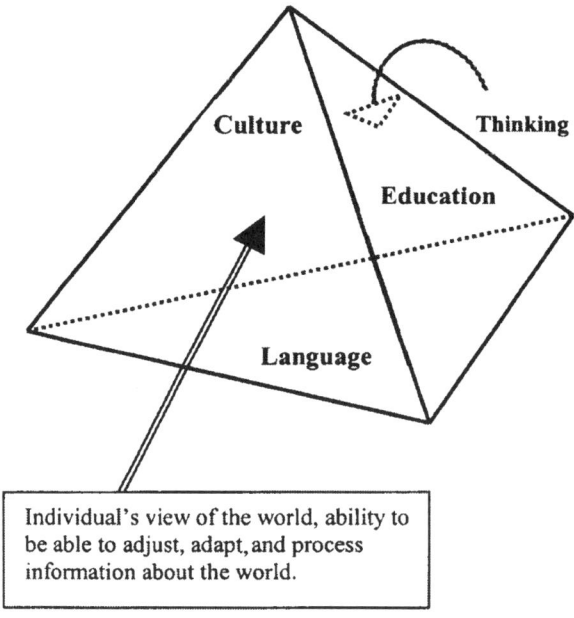

Figure 2. The prism model.

Through ICT based activities built upon specified language objectives, students should be provided with opportunities that facilitate the acquisition of English and the development of an English communicative proficiency, and these need to be the primary objectives of CALL classes. Although research is required to confirm that interaction and language use in and through ICT will transfer to face-to-face communicative and linguistic proficiency, the above process may relieve confusion between the computer and language learning components of a CALL class. Reducing these peripheral issues will enable the language components of CALL to be more evident, and hopefully, to be utilised by the students. Secondary objectives – the development of ICT

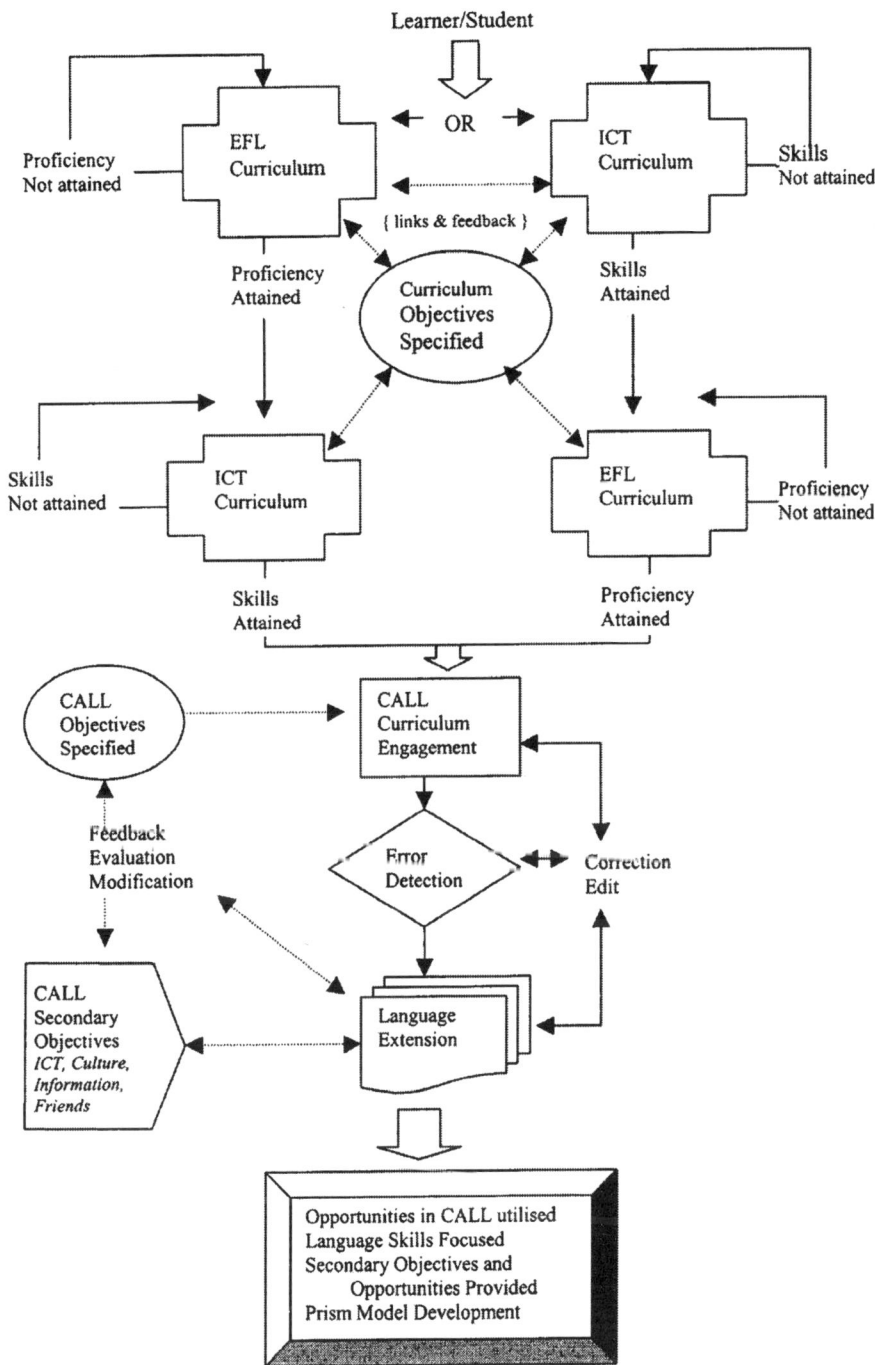

Figure 3. The theory of learner development.

skills, a greater world-view and cultural awareness, and a sense of personal and educational achievement – may also be realised. Further research could extend this research on the influences of CALL and ICT on Japanese culture and language and interactions therein.

7. A theory of learner development

The theory of curriculum development (Fig. 3) highlighted the need to develop ICT skills in the Japanese language and improve English language proficiency before participating in a CALL class. In the theory of learner development the interactive processes of these courses are developed. It is essential to understand that a student's participation in the ICT Skills and EFL proficiency courses do not necessarily have to occur sequentially; a student may undertake both courses simultaneously and this may be the optimal method. Nevertheless, a student choosing to study EFL in a CALL class should undertake the prerequisite courses until the proficiency or skill is achieved. It is important that students have a relatively equal starting position in the CALL based curriculum so that the major focus can be on language acquisition and production and not on computer skill development.

Although the ICT skills and the EFL proficiency courses are separate and distinct, there need to be links and feedback between the two from students and by teachers of the courses. This feedback and evaluation will assist teachers to (a) develop the courses, (b) alleviate confusion about the objectives and goals of the course/s, (c) facilitate student learning choices and strategies, and (d) advise students about the selection of future academic courses and strands. Linking the two may also assist students to build schemata (ICT skills, EFL proficiency, and an ICT-EFL relationship) which may assist with their future language and academic studies.

When choosing to continue EFL studies through a CALL curriculum, and if the objectives of the curriculum are specified, students may be better able to make learning choices and utilize opportunities provided in CALL classes. As previously outlined, the primary objective of a CALL-based curriculum should be the acquisition and production of the target language, not the development of ICT skills, and this is the expected and desired outcome expressed by students. The secondary objectives, such as ICT skill development, increased culture awareness, opportunities to gather new and different information, making of new (international, local, online, offline) friends, also need to be specified and are important factors in the process of learning a second or foreign language.

If the primary objective of a class is not language acquisition, the acronym CALL should not be applied, but instead the terms CAI (computer assisted/aided instruction) or CAL (computer assisted/aided learning). CALL pedagogical philosophy is language learning, whereas, the pedagogical philosophy for CAL/CAI can be applied to any field of study. Although this may seem to be a discussion of semantics, clarity of definition in understanding the philosophy of CALL will enhance effective and clear objectives from which students can develop learning strategies and choices of academic study.

The provision of opportunities to learn and produce language alone will not, however, produce language or communicative proficiency. Learners need to be made aware of correct language forms during production, and errors need to be recognized – particularly errors that impair communication so significantly that they distract the listener or reader from the message in the students' speech or writing (Hendrickson, 1987).

Correction and editing of errors need not be left to teacher input only. ICT provides opportunities for the learner to evaluate his or her own English language production using spelling and grammar checkers. ICT, furthermore, is interactive, involving the computer and often a second person. This person, whether a peer or anonymous, can provide useful correction and feedback in the communicative process.

It is difficult, however, to expect another person to offer language error-correction advice during a communicative event. The difficulty of correcting student errors without extinguishing communication and production during a speech act or communicative event is a difficult task the teacher encounters regularly. It may, therefore, be even more difficult for a peer or non-native speaker. Nevertheless, language production can be recorded by ICT and interlocutors should be encouraged to evaluate their speech acts in the communicative event retrospectively. In other words, written interaction can be evaluated and corrected (with or without teacher guidance) by students.

Through language extension activities, students are able to utilize language skills in focus through the use of, or engagement with, the secondary objectives. If opportunities in CALL are utilized, it is hoped that a higher proficiency in the use of English by the Japanese university student will subsequently result. The hope that there is transference from the language produced and learned in a CALL based curriculum to a face-to-face communicative event should be high priority research supported by educational and financial institutions. Research is required to substantiate and justify the use of ICT in a CALL based curriculum for EFL education in Japan.

8. Summary

This paper has provided two theoretical models for the use of a CALL based curriculum at university level EFL studies in Japan. These were developed from student comments posted to a web-based bulletin board. Comments revealed that students were confused about the objectives of a CALL class, particularly over the dichotomy between the ICT skills nature of the classes and their EFL education component. This confusion, combined with different ICT skill and language proficiency levels amongst students in any one class, created conflicting pressures on learning repertoires.

The two models attempt to provide a process that will alleviate the conflicts expressed by the students. The first model (theory of curriculum development) attempts to focus on curriculum development processes, whereas the theory of learner development proposes a process to facilitate learner development in both English language proficiency and ICT skill development. These processes are needed for the Japanese language learner, as the individual's exposure to ICT and all-English EFL classes is often limited. Japanese university students' schemata need to be broadened by the development of ICT skills in their first language and through all-English EFL courses. Furthermore, CALL curriculum and pedagogical objectives should be explicit so students can make informed choices and utilize the language and educational opportunities made available through ICT in a CALL based EFL university programme.

Note

1. Modification of the student postings has been kept to a minimal. Amendments in English to assist with meaning have been closed in [square brackets]. Japanese translations have been placed in (regular brackets).

References

Bachnik, J. M. (1999). Do IT yourself: Assessing the information "revolution" in Japanese higher education, *National Institute of Multimedia Education's International Symposium, 1999, Japan*. Retrieved from www.nime.ac.jp/conf99/pre/Bachnik.html

Bannan-Ritland, B., Harvey, D. M., Milheim, W. D. (1998). A general framework for the development of Web-based instruction. *Educational Media International, 35*(2), 77-81.

Bennett, F. (1999). Education and the future. In *Educational Technology & Society, 2* (1). Retrieved from zeus.gmd.de/ifets/periodical/vol_1_99/fbennett_short_article.html

Blair, M. (1996, October). *A multitheoretical analysis of the impact of information technology on higher education*. Paper presented at the 10th Annual National Conference on Liberal Arts and the Education of Artists, New York.

Chapelle, C. A. (2001). *Computer applications in second language acquisition*. Cambridge: CUP.

Cuban, L. (1986). *Teachers and machines: The classroom use of technology since 1920*. New York: Teachers College Press.

Hendrickson, J. E. (1987). Error correction in foreign language teaching: Recent Theory, Research and practice. In M. H. Long & J. C. Richards (Eds.), *Methodology in TESOL: A book of readings* (pp. 355-369). Boston, MA: Heinle & Heinle.

Jones, A., & Mercer, N. (1993). Theories of learning and information technology. In P. Scrimshaw (Ed.), *Language, classrooms, and computers*, London & New York: Routledge

Knobel, M., Lankshear, C., Honan, E., & Crawford, J. (1998). The wired world of second-language education. In I. Snyder (Ed.), *Page to screen: Taking literacy into the electronic era* (20-50). London: Routledge.

Kuramoto, A. (1999). A cognitive psychological study of L2 learning environments. In P. N. D. Lewis (Ed.), *Calling Asia: The proceedings of the 4th Annual JALT CALL SIG Conference, Kyoto, Japan, May 1999* (115-118). Nagoya: Chubu Nihon Kyouiku Bunkakai.

Laurillard, D., & Marullo, G. (1993). Computer based approaches to second language learning. In P. Scrimshaw (Ed.), *Language, classrooms, and computers* (143-165). London & New York: Routledge.

Marjanovic, O. (1999). Learning and teaching in a synchronous collaborative environment. *Journal of Computer Assisted Learning, 15*(2), 129-138.

McMahon, J., Gardner, J., Gray, C., & Mulhern, G. (1999). Barriers to student computer usage: Staff and student perceptions, *Journal of Computer Assisted Learning, 15*(4), 302-311.

Narita, M. (1999). Barriers for educational use of the Internet in Japanese higher education, *National Institute of Multimedia Education's International Symposium, 1999.* Retrieved from Japan www.nime.ac.jp/conf99/pre/Narita-Masahiro.html

Nelson, G. (1998). Internet/Web-based instruction and multiple intelligences. *Educational Media International, 35*(2), 90-94.

Panel unveils 5-yr IT strategy plan. (2000, July 7). *Yomiuri Shimbun.* Available from www.yomiuri.co.jp/index-e.htm

Sakamoto, T. (1992). Impact of informatics on school education systems: National strategies for the introduction of Informatics into schools – Nonsystematic but still systematic. *Education and Computing, 8,* 129-135.

Selwyn, N. (1998). The effect of using a home computer on students' educational use of IT. *Computers and Education, 31,* 211-227.

Selwyn, N. (1999). Technological utopianism and the future (in)perfect: A response to Fred Bennett. *Educational Technology & Society, 2*(1). Retrieved November 1999, from zeus.gmd.de/ifets/periodical/vol_1_99/nselwyn_short_article.html

Sloane, A. (1997). Learning with the Web: Experience of using the World Wide Web in a learning environment, *Computers and Education, 28*(4), 207-212.

Sugimoto, T. (1999). Three critical "gaps": Successes and failures of computer literacy in Japanese education, *National Institute of Multimedia Education's International Symposium, 1999.* Retrieved from www.nime.ac.jp/conf99/ pre/ Sugimoto.html

Van Dusen, G. C. (1997). The virtual campus: Technology and reform in higher education. *ASHE-Eric Higher Education Report, 2*(5). Washington DC: The George Washington University.

Vygotsky, L. S. (1978). *Mind in society.* Cambridge, MA: Harvard University Press.

Warschauer, M. (1998). Online learning in sociocultural context [Electronic version]. *Anthropology & Education, 29*(1), 68-88. Retrieved September, 2000, from www.gse.uci.edu/markw/elcc intro.html

Whorf, B. (1940). *Science and linguistics, language thought and reality.* Cambridge, MA: MIT Press.

Whorf, B. (1956). *Language thought and reality.* Cambridge, MA: MIT Press. Quoted from Musgrave, F. (1982). *Education and anthropology: Other cultures and the teacher.* New York: Wiley and Son.

2

Applying Hypertext Concepts to Language Acquisition in a CALL Environment in Japan

Kevin Ryan
Showa Women's University

1. Introduction

Hypertext is more than the WWW; its surrounding concepts were developed long before Tim Berners-Lee put finger to keyboard. Concepts such as linking, data structuring, document standards, survivability, semantic mapping, and data warehousing all have enormous potential when applied to language learning and teaching. This paper will briefly explore the theory behind hypertext, look at the state of the art of hypertext use in Japan through examples, and then propose different ways to apply the current and future hypertext technologies to increasing the rate of language acquisition.

1.1 The path of hypertext

Roosevelt's Science Officer wrote an article (Bush, 1945) that explained the need for a Memex, a device that made connections between bits of data. The irascible but tenacious Ted Nelson expanded on this idea in his book *Literary Machines* (1980), which describes his legendary project Xanadu, an initiative toward a universal electronic literature, proposing to link all human knowledge. Concurrently, technical advances at IBM led to a standard for codifying information, called Standard General Markup Language (SGML), the grandmother of HTML and the newer XML. Peripheral but important advancements with neural networks in artificial intelligence (AI) have led to ideas about semantic networks (Jubak, 1992), a cross between a database and hypertext. The development of the World Wide Web by Tim Berners-Lee and what has followed is relatively common knowledge.

Ideas about applying hypertext to language and literature began as a solution to the programmed learning approach, seen to lack motivational and pedagogical validity. George Landow (1997) at Johns Hopkins University applied the ideas of French philosophers such as (deconstructionist) Jacques Derrida in developing concepts of hypertext in literary criticism such as blurring the roles of reader and writer, nonlinearity, and multilinearity of texts and metadocuments. See *A Subjective Chronology of Literary Hypertext* by Stuart Moulthroup (2000) for a more historical perspective.

2. Elements of hypertext

The four basic elements of hypertext are the node, anchor, link, and target. The node is most often text, but can be any type of multimedia. "Multimedia in a CALL environment means that input from written text may be enhanced by pictures, graphics, animations, video, and sound as well as hyperlinks to other explanatory texts" (Hanson-Smith, 1999, p. 189). Text nodes can be from one word to hundreds of pages long. Each node is an idea or collection of ideas. An anchor is the point from which a link is taken, that link ending at the target. Both anchors and targets are parts of nodes, and can also vary in size and importance.

Hypertext contains methods of navigation through these links, but should also contain ways for the reader to add comments or additional information. Thus the roles of reader and writer are blurred. Robert Wachman (1999) writes that multimedia authoring software should include the powerful tool of hypertext links to add non-linearity to programs and lessons. He also advocates empowering the student to add comments and make links in any such material created with authoring programs. Hanson-Smith provides an example: "In HyperStory . . . hyperlinks point out metaphors and symbols and provide study questions and ideas for writing assignments. When a hyperlink is accessed, a simple word processor allows students to take notes and begin composing an essay" (1999, p. 203).

Links are the core of the hypertext system, and this element can be used to either enhance the openness of the hypertext by linking to nodes outside the proprietary hypertext, or to constrain the size of the domain. We will use the definition of domain here in its artificial intelligence (AI) sense: the entire range of content material covered. Links are usually seen as open-ended, but in language learning, where the cognitive load of 1) learning how to navigate a hypertext, 2) learning how to operate a computer, and 3) learning the target language, constricting the domain is usually recommended. "Links do not only express semic relations but also, significantly, establish pathways of possible movement within the Web space; they suggest relations, but also *control access to information*" (Burbules, 1997, p. 105). Burbules goes on to note that in many cases the link itself will become invisible, altering the document without perception by the reader. Indeed, in the communicative approach, restricting the domain is a key tactic for beginning and intermediate students.

Shapiro found that "when the learner has no prior knowledge [of the content domain], the influence of an overview is powerful enough not only to guide the structures of a novice's internal representations, but to overshadow the effect of the learning goal during the process" (2000, p. 58). This overview, similar to a table of contents, is the main element of hypertext organisation, or at least the one most often used by learners. This organisation, if matched to the task of the learner, worked well. But if the goal of learner and organisation of context domain were at odds with one another, the organisation of hypertext had the greater effect on learning, weakening the results of the conflicting task.

Indeed, our students may actually perform better using hypertext interaction if they are of a younger generation (*sensei*, the word for teacher in Japanese, literally means "born before"). Currently, we are in a transition stage of processing information, of which language ability is an integral part. Somewhat upsetting, the following quote nonetheless predicts what will most likely happen as information becomes a more central part of our lives. Hypertext is certainly one of the new ways that this processing of information occurs, but as the shift to the *Age of mind* (see below) occurs, hypertext and language learning will have to be embedded within each other:

> Laurence Heilprin (1989), a noted American information scientist has characterised the emergence of the *information society* in two phases, each with a different persona and locus of scholarly concern. The first phase, labelled the *Age of information*, describes the exponential growth, organisation, and global distribution of information. Heilprin asserts that during this phase, the generation, processing, provision, and use of information has emerged as the predominant human activity, and much scholarly energy has been spent in developing laws, theories, generalizations, policies, and speculations about these. However, he identifies a fundamental shift in our thinking about the information society, labelled as the *Age of mind* where the emerging focus appears to have shifted from the phenomenon of information itself to understanding people as human information processors; and how this understanding can inform the effective organisation and provision of information in society. (Todd, 2000, p. 103)

Japanese students are notorious for their inability or unwillingness to ask questions. Hypertext can be used to overcome this by organising information in such a way that questioning is the only way to complete activities:

> Questioning is the basis for information literacy – the ability to interpret information and extract or create new meaning, to solve problems, and to make decisions based on reliable evidence and a thorough understanding. Questions enable us to search for pertinent information and convert data into information and information into insight. (McLean, 1999, p. 93)

Action mazes (see below) are one type of hypertext fiction used in language learning as a stimulus for question-posing in discussion.

Building a hypertext for learning, and particularly language learning, requires deft management of these elements. One way to consolidate hypertext into something students can easily grasp is to anthropomorphise it into a *tutor*. Ohmaye (1998) notes that a good learning simulation does not simply immerse the learner in content. A virtual tutor is an indispensable tool when designed properly. A tutor is best used as a source of information under an apprenticeship model of learning, and (echoing the dichotomy of *magister* and *pedagogue* developed by Higgins; see below) should not correct grammar. Ohmaye notes that grammar correction does not always lead to improvement. A good tutor provides constructive feedback, but corrects learner input only when semantically inappropriate. That feedback may contain physically or conceptually accurate forms, or both. The latter is more faithful to the underlying concepts of the language being taught, rather than the surface language itself. A good tutor also provides a means for interaction. Truly personalised feedback must take into account both motivational and cognitive situations of the learner at that moment.

3. Qualities of hypertext

Hypertext can take many forms, but there are qualities common to all. Some can be more readily applied to language learning. Unfortunately, research into the effectiveness of hypertext on language learning is still in the exploratory stage. Ayersman (1996, p. 503) notes that initial hypermedia research was overwhelmingly attitudinal and perceptual in nature. Later strands included individual user differences, systems analyses of the software, and effects on the "cognitive and procedural performance" [learning] of the user.

Altun (2000) reviews the literature on differences between linear and nonlinear (hyper) text and finds that there may *not* be a fundamental difference between printed and electronic texts. He also finds that "hypertext readers tend to transfer their printed text reading strategies to computer environments" (2000, p. 38). Overall, further research needs to be conducted on how links and hypertext affect the average Japanese student. Motivation as well as information access are primary considerations here.

Altun (2000) studied two staff members at a university computer centre, both native speakers of English with more than ten years of computing experience. Even these experienced users noticed that they did not follow links unless they found them appealing or attractive. Both were annoyed at the time lag between link selection and display (as are many of us with slow Internet connections), finding that poor page display caused disorientation. The expert users said they did not usually bother with glossary items, but results show they often did click on links to definitions. Both lamented the fact that they had to use a pen to jot down pertinent information, instead of noting it directly into the hypertext (p. 39).

4. Hypertext and language learning

John Higgins makes a wonderful distinction of the teacher's role in language learning for the computer that can be easily applied to hypertext (noted above). His 1984 article *Learning with a Computer* explains that computers had been used mostly in the role of *magister*, a taskmaster that gave lectures, quizzed students, and kept records, much as the medieval German professors the name refers to. He proposes another model, the *pedagogue*, a servant in India hired by the rich to act as a resource for their children, always available to pose interesting questions or supply answers. Higgins found the computer as pedagogue is more effective for language learning.

Mario Rinvolucri's adaptation of a business training technique called *action mazes* in the early 1990s led to the first true application of hypertext in language teaching. In his (sadly, out-of-print) book of the same name, he uses a problem-posing approach to coax natural language out of students by having them focus on an interactive short fiction. This kind of fiction is often called *branching hypertext*. When the baby starts crying in the middle of the night, or a neighbour continually parks in your space, there are options to consider. Each option leads to another choice and another. Students are encouraged to discuss ramifications along the way. Most outcomes are negative, forcing students to backtrack and redefine priorities, so that a short hypertext (20 nodes with 60-80 links) may require an hour or more of intense discussion. Students are motivated by the fact that they must understand each node and its links before making a choice, and that their choice will have consequences.

Software applications for creating hypertext are too numerous to review here, but two deserve special mention. The first of these is HalfBaked Software's Quandary (Arneil & Holmes, 2000). Written by two ESL teachers turned developers, Quandary is an authoring tool that greatly simplifies writing and creating a hypertext. Building the action maze on paper requires intensive planning beforehand. With Quandary, one can begin with a rough outline and fill in the holes as one builds the hypertext. Quandary keeps track of the connections to each node so that one can begin with a relatively simple structure to the hypertext and add layers of story to the action maze. More on constructing a hypertext follows.

Wachman (1999) exhibits the use of a program called HyperCard where students create a branching story whose outcome is determined by choices along the way. Putting the hyperlinking tools in student hands can be beneficial for critical thinking as well as language use. This is the goal of most hypertext-literate teachers in Japan: to have the students use a hypertext application to organise their thoughts.

5. Hypertext and the four skills

In dealing with links, we may find that students with little traditional reading and writing skills may fare even better than students steeped in paper-based skills. Using an electronic text with hypertext requires different skills.

> Faced with the differences between p-texts and e-texts and the challenges of hypertext, users of computers need to adopt new reading skills and even new habits of thought. Although the transfer of skills from paper-based texts can be useful to a certain degree, with skilled writers using word processing more effectively (Corbel, 1997), the old skills may ultimately get in the way of effective work with electronic texts. Linear, printed text supports critical thinking in which the text is considered holistically . . . . Hypertext, in contrast, supports associative thinking, encouraging the reader to move between initially unrelated pieces of information linking separate, interconnected concepts in ways that are not linear. (Thurstun, 2000, p. 69)

Kol and Scholnik (2000) studied screen-reading strategies and had to redefine the terms skimming and scanning to suit screen reading. Screen scanning was defined as "quickly searching for specific pieces of information by using the Find feature of the word processor" (p. 70). They defined screen skimming as "reading the hyperlinked outline provided, clicking the outline to access specific sections of the text, quickly reading and highlighting those sections, and scrolling to read the highlighted sections to get the main ideas" (p. 70). Results showed that "students reading from the screen and students reading from paper performed equally well in all types of questions on the reading comprehension test" (p. 74).

Hanson-Smith (1999, p. 203) also points out the economics of using hyperlinks for vocabulary study, allowing the student almost instant access to explanatory nodes. Software packages allow for limiting access to these links so as not to provide too great a cognitive load. Hanson-Smith goes on to advocate teacher guidance and control of hypertexts, encouraging prediction and other useful strategies along the way:

> However, hypermedia may also prove a fatal distraction when students fail to use prediction strategies at all or when they follow hyperlinks in several different directions and lose their way back to the original text. (Hanson-Smith, 1999, p. 203)

Even though roles of reader and writer are blurred in hypertext, language learners can benefit from applying hypertext concepts to common word processing programs:

> Software like Daedalus Integrated Writing Environment (DIWE, 1997), CommonSpace (1997), and Web-B-Mail (Pfaff-Harris, 1996) allows students to form small groups electronically (either within one classroom on a local area network [LAN] or across the world on the Internet) in order to brainstorm ideas, read and respond to each others' writing with comments in margin-windows, and assist in peer-editing and evaluation. (Hanson-Smith, 1999, p. 210)

As we can see, adding features that allow for cross-reference, access to additional information, and contact with other writers affords L2 writers more resources at their fingertips. It is like having that pedagogue at one's side when attempting to write in a second language, constantly adding and gently suggesting improvements.

Adding peer and instructor connections to any text also allows for additional language practice that proves to be an additional benefit. "The email messages of second language learners do provide other learners with grammatical, targetlike input displaying a range of language features similar to those used by first language speakers of English in comparable genres" (Holliday, 1999, p. 187). Students get authentic interaction with comprehensible input on the task at hand, writing a text in English.

6. Current uses of hypertext in Japan

6.1 eigoTown.com

The web-based English language learning facility eigoTown.com is a prime example of hypertext use in a double-byte EFL environment. While there are larger ELT sites on the Internet, none with native speaker management focuses specifically on Japanese learners. This necessitates some adaptation to the Japanese market absent in sites such as EnglishTown.com or Global English. The eigoTown.com website caters to the false beginner busy at work or school, wanting to pick up English during free moments throughout the day. Thus the nodes must be relatively short (and interesting) with numerous links among the sections of the site. The site contains sections called eigoCollege, Study Abroad Plaza, Culture Café, Community (message boards with ePals connections), a job centre, a store for learning materials, and a school search. As is clear from the last item, eigoTown aims to complement traditional English language study with a great body of topical information easily accessed and digested. For lower level learners there is information in Japanese about English language and cultures of English speaking countries. While it looks initially like the web-based English study magazine common in Japan, with most content in Japanese, lots of gloss, but few unusual vocabulary items, hypertext makes the difference.

Because the site was set up so carefully (and is continually tweaked according to reader feedback), it acts very differently from a magazine. Although much of the site appears to be in Japanese, intermediate and advanced users can navigate quickly to content challenging for them. Japan has had a long tradition of studying about English in Japanese, but this notion is turned on its head at eigoTown.com. One example is the mouseover; when the pointer is on a section tab, the title switches from Japanese to English. Readers (users) of the site access information quickly via the many links through and among different sections. Navigation paths are clearly marked so that each user gets what he or she wants. The contents appeal to the largest groups of English learners in Japan, high school and university students, businesspeople, and hobby students. Some sections help improve scores on the all-important qualifying exams given for English across the nation such as TOEIC, TOEFL, and STEP. Resources such as dictionaries pop up wherever needed. The eigoTown site acts as a doorway to the English world of words and ideas through a friendly, interesting, and easily accessible interface.

6.2 StorySpace for comparative cultures

M. Antoni J. Uceler (personal communication) uses hypertext both in class for lectures and for student projects. Using Eastgate's StorySpace (1990-2001), he is able to bring the wide variety of topics covered in the Faculty of Comparative Culture at Sophia University in Tokyo to his students, who are from all over the world, with about half from Japan. He lectures on philosophical and intellectual differences between east and west, most specifically connections between Japan and Europe in the 1600s.

This reflects the perceptions of hypertext users in other countries. "Many educators are adopting hypermedia as a construction tool for collaboratively creating their own software suited to their particular disciplinary needs" (Ayersman, 1996, p. 501).

Lecture notes displayed in class with many visual cues such as colour-coding and placement on the screen benefit all students, but especially Japanese students. There is likely some correlation between their success and their being quite field dependent.

Using hypertext lecture notes provides Ucerler with flexibility in topics covered. If a student asks a related question, the answer, with appropriate references, is one click away. Ucerler also builds his lecture notes while in class. If a subject is not covered, he adds notes on the new topic to the hypertext during his lecture, giving students an idea of where in the subject domain the new sub-topic exists. Between classes he fills in those notes with appropriate references. The hypertext is available for student reference on his LAN at all times, and is continually updated with these new sub-topics. In a way, these lecture notes become a flexible core of knowledge for the entire course.

According to Ucerler, student behaviour changes when lectures are given with hypertext. Students tend to ask more questions. Also, more students ask questions, with those less capable taking a more active role than in traditional lectures. Students report that they feel more relaxed in a hypertext lecture. The hypertext allows for scalability to different levels of both content and language, important in a class with both native and non-native speakers with varied educational backgrounds.

Ucerler has been experimenting with combinations of hypertext with other media. He often gives students traditional lecture notes in English, and displays Chinese or Japanese characters on screen with their referents to clarify his multilingual subject. He has even incorporated a PowerPoint presentation, which has basic hypertext capability, but works better in linear "outline" fashion, as a presentation tool, with the reference and follow-up material in StorySpace hypertext. He switches back and forth between linear and hypertext modes of presentation as necessary during his lecture. StorySpace can be used as a kind of whiteboard, linking historical and philosophical subjects to contemporary political ones, for example.

Ucerler adds that StorySpace has changed the way he organises knowledge, leaving the traditional outline for a quickly created hypertext, with many nodes that can later be linked in multiple overlapping organisational frameworks. Hypertexts are never finished, allowing continual expansion, augmentation, and enhancement.

Writing hypertext lecture notes creates a need for organisation on one level. Ucerler has found that for the overall course, a navigation tool is essential, displayed at all times during lectures to give students a schema for the domain. The hypertext also allows for flexibility within this organisation – if appropriate, links can be inserted in the nodes of information to create multi-layered organisation schemes.

Changing the hypertext from a reading to a writing space is one direction Ucerler intends to go in the following semester. Students will contribute information to the hypertext directly and expand class notes at a much higher rate, further involving themselves in the creation of the domain. Collaborative efforts of this type prepare students for a research environment after they graduate. Ucerler's use of hypertext in class prompted a number of students to change the form of their final papers to a PowerPoint presentation, which has rudimentary hypertext tools and is easier to use than StorySpace.

Some undergraduates cannot wait to do hypertext versions of research projects. One student in Independent Studies says that writing in a hypertext environment helped her paragraph organisation tremendously (nodes of information are often the size of paragraphs). She was able to look at her controversial subject (an embassy of four boys sent by the *daimyo* from Japan to Europe) from a variety of viewpoints, thus enhancing critical thinking. The graduate student has moved from a classic thesis-antithesis-synthesis interpretation to give freedom of interpretation to the reader, presenting conflicting viewpoints almost simultaneously. The graduate student based her hypertext writing style on how Ucerler presented his class notes. She emphasized the intertextuality of Derrida and Barthes, by alluding, contradicting, and supporting others' ideas within her hypertext.

7. Writing with hypertext

A semester-long project in the author's own classes, with a final goal of creating a hypertext, aims at teaching students about the expanded horizons which using a computer as a writing tool allows. Using StorySpace as a language-learning tool is uncommon in both Japan and other parts of the world. "Hypermedia can be used in the classroom in several ways. Perhaps the approach used least often, but thought of most frequently, is hypermedia as a standalone commercially available software program" (Ayersman, 1996, p. 501). However, hypertext software use, especially for putting writing projects on the web, is accelerating as software gets easier to use. Many website authoring tools such as Macromedia's DreamWeaver and Microsoft's FrontPage are approaching the utility and ease of use that StorySpace has enjoyed for years.

In the first session of the author's introductory course (Writing with Computers), English language major students view past projects. They are astounded to find that they, in the space of 12 weeks of 90-minute classes, will learn to use the computer and word-processing software as a writing tool. To accomplish that they have to learn typing (a necessary evil), and develop computer literacy. The middle part of the semester coincides with their normal writing classes in that they study how to write paragraphs and put those paragraphs into an essay. While greater latitude is taken in the normal writing course, a formulaic approach minimizes the mechanics of writing with computers.

We concentrate on breaking down a large hypertext topic between classes (five sections in the fall semester, totalling just over 100 students), "chapter" subtopics broken down into group "section" topics, which are broken down into individual topics. However, it does not stop there. Individual topics are broken down into paragraph topics, and from there into sentence topics. Thus each sentence in each paragraph in each essay in each section in each chapter in the hypertext is a branch from the "source" topic. Students see the relation of their work to the whole.

In the last week of class, students submit their essays and the author enters them into StorySpace using a branching structure. The best part comes on the final day, when students, with their final five-paragraph essays (three paragraph topics with introduction and conclusion) link the content of their sentences to those of others in their group (1 point for each), in their class (2 points), and to ideas in other classes (3 points). This creates pandemonium and panic at first, but soon students realize how well-organised the branches are, and how easy it is for them to skim and scan the content to find relationships among sentences. The author leaves the class with a sheaf of links that make the branch structure into a riotous web of phonological, semantic, linguistic, and conceptual links built by students during the final test. It takes days to enter all the links. In more advanced classes, students take over and manage StorySpace itself. As expected, they have not reached the level of Ucerler's students. However, it is noteworthy that the English used to discuss the structure of the hypertext document is often richer than the topic itself!

8. Building hypertext

Building a hypertext is fraught with complexities unless there is focus. Action mazes and fiction are particularly hard to constrain. Nonfiction hypertexts are often easier to build, but both types can spiral out of control because the linking process leads to exponential growth if each option in turn is expanded upon. Figures 1 to 5 show some graphical representations of ways of organising hypertexts to eliminate excess (except for the last, which is an example of what not to do).

8.1 Five hypertext structures

The other most common error of novice hypertext writers is to leave nodes dangling, with only one reference, or perhaps none. Building a kind of multidimensional table of contents is necessary to use as a navigational map of the site or hypertext. This takes time but can be simplified using authoring programs such as Quandary and StorySpace. The latter has three different representations of nodes and links, for example.

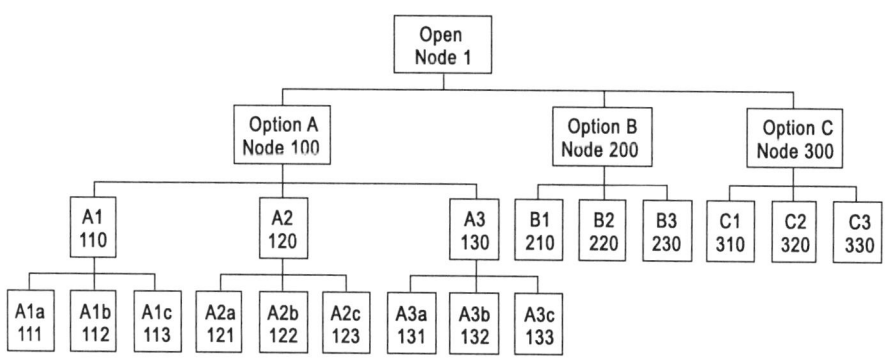

Figure 1. The three-level pyramid.

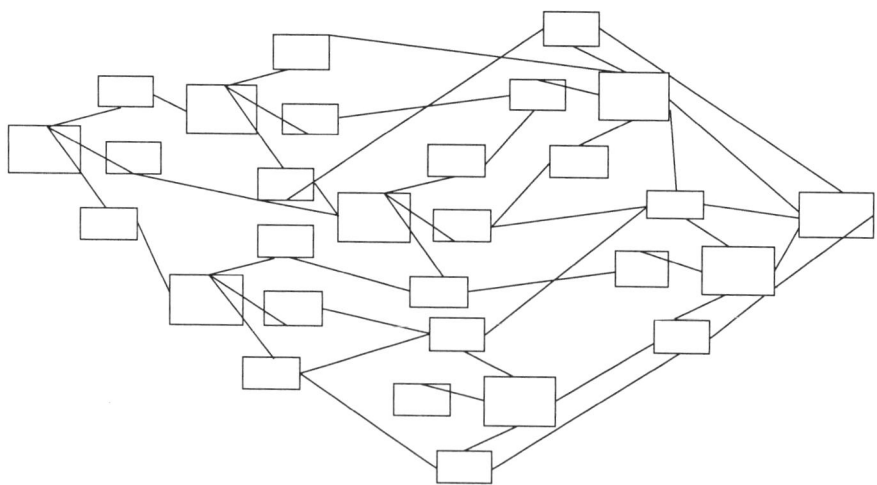

Figure 2. The blob.

Adding some kind of search mechanism is essential too. Learning styles dictate how much a hypertext is used, so it must be flexible enough to cater to all. In building a hypermedia learning system, Tergan and Lechner (2000) found that harmonizing information seeking, browsing, and searching were important "because they provide functions which complement each other during processes of information localization and learning" (p. 213). To clarify the differences of these processes an example of a bookstore might help. Browsing might be wandering the aisles, seeking would be asking the clerk for a good spy novel, and searching could be likened to giving the clerk an ISBN number.

Kashihara et al. (2000) believe that exploring content domains is an effective way of learning, and that this exploration is "a self-directed and constructive mental activity" (p. 254). They have developed a system of support for exploratory learning that

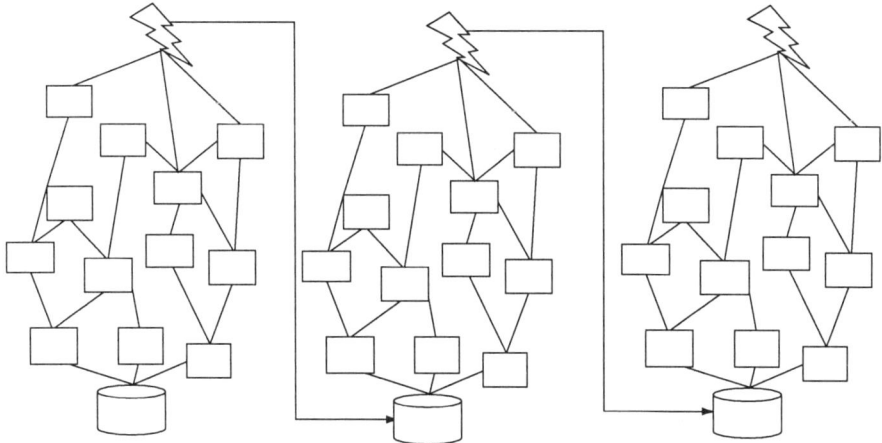

Figure 3. The chemistry set.

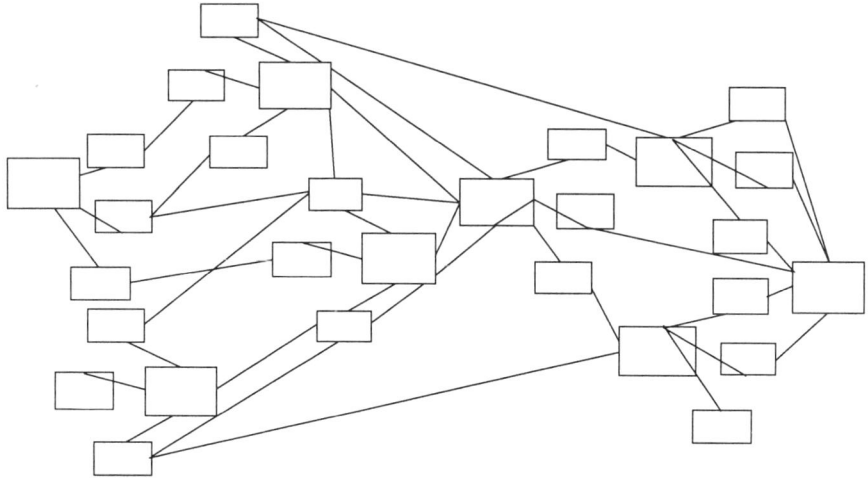

Figure 4. The hourglass

either presents learners with as many options as possible or, conversely, limits exploration space allowing it to expand only gradually. The space is limited by control levels; embedded information structure, limiting information resources, limiting exploration paths. These all lead to design considerations.

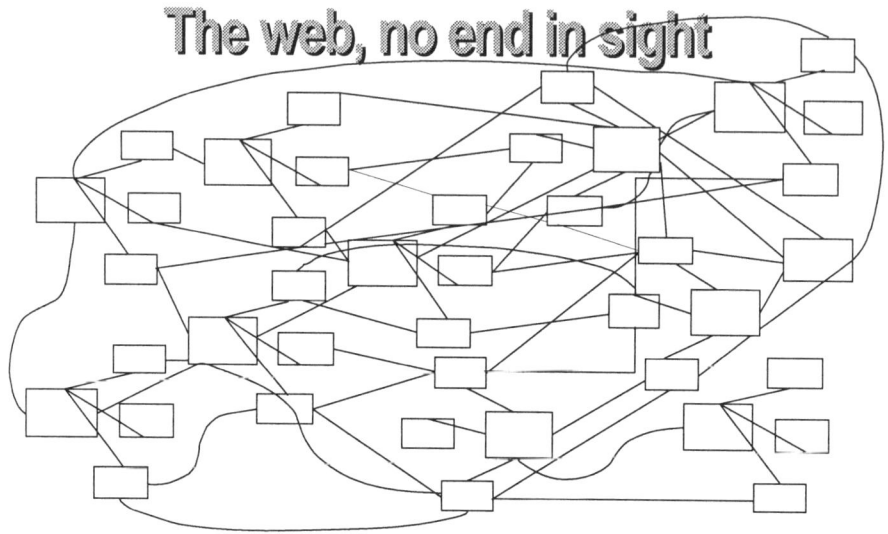

Figure 5. The Web.

The future brings with it more ease of use. Visualization of the hypertext in multi-dimensions, speech recognition, and database integration are all on the horizon. One technology already here, which should prove most formidable, is a common basis for tagging elements within a discipline so that not only can authors link their databases, but machines can "talk to each other" and create a huge hypertext on a subject on the fly, immediately, and globally. The technology, XML, is ready. We only need to agree on semantic standards to bridge knowledge into hypertext to finally create Bush's Memex or Nelson's Xanadu (1990). I cannot wait.

Disclosure

The author has just begun working for ELT News, a web-based language teaching magazine with eigoTown.com, the parent company (2001).

References

Altun, A. (2000). Patterns in cognitive processes and strategies in hypertext reading: A case study of two experienced computer users. *Journal of Educational Multimedia and Hypermedia, 9*(1), 35-55.

Arneil, S., & Holmes, M. (2000). Quandary [Computer software]. Retrieved October 2001, from www.halfbakedsoftware.com/quandary/

Ayersman, D. J. (1996). Reviewing the research on hypermedia-based learning. *Journal of Research on Computing in Education, 28*(4), 500-525.

Bush, V. (1945). As we may think. *The Atlantic Monthly, 176*(1), 101-108.

Eastgate (1990-2001). *StorySpace.* Retrieved October, 2001, from www.eastgate.com/storyspace/StoryspaceOV.html

EigoTown.com (2001). Retrieved October, 2001, from www.eigotown.com

Halliday, L. (1999). Theory and research: Input, interaction and CALL. In *CALL Environments: Research, practice, and critical Issues* (181-188). Alexandria, VA: TESOL.

Hanson-Smith, E. (1999). Classroom practice: Using multimedia for input and interaction in CALL environments. In *CALL environments: Research, practice and critical issues* (189-215). Alexandria, VA: TESOL.

Higgins, J. (1984). *Language, learners and computers.* London: Longman.

Jubak, J. (1992). *In the image of the brain: Breaking the barrier between the human mind and intelligent machines.* Boston: Little & Brown.

Kashihara, A., Opperman, R., Kinshuk, Rashev, R., & Simm, H. (2000). A cognitive load reduction approach to exploratory learning and its application to an interactive simulation-based learning system. *Journal of Educational Multimedia and Hypermedia, 9*(3), 253-276.

Kol, S., & Schcolnik, M. (2000). Enhancing screen reading strategies. *CALICO Journal, 18*(1), 67-80.

Landow, G. P. (1997). *Hypertext 2.0: The convergence of contemporary critical theory and technology.* Baltimore, MD: Johns Hopkins Press.

Molthroup, S. (2000, June 8). *A subjective chronology of literary hypertext.* Retrieved from raven.ubalt.edu/staff/moulthrop/chrono.html

Nelson, T. (1990). *Literary machines: The report on, and of, Project Xanadu concerning word processing, electronic publishing, hypertext, thinkertoys, tomorrow's intellectual revolution, and certain other topics including knowledge, education and freedom* (revised edition). Sausalito, CA: Mindful Press.

Ohmaye, E. (1998). Simulation-based language learning: An architecture and a multimedia authoring tool. In *Inside multi-media case-based instruction* (1-101). Mahwah, NJ: Lawrence Erlbaum.

Shapiro, A. M. (2000). The effect of interactive overviews on the development of conceptual structure in novices learning from hypermedia. *Journal of Educational Multimedia and Hypermedia, 9*(1), 57-78.

Tergan, S. O., & Lechner, M. (2000). HyperDisc: A hypermedia-information system for supporting advanced studying design rationale, application, and some research results. *Journal of Educational Multimedia and Hypermedia, 9*(3), 207-222.

Thurstun, J. (2000). Screenreading: Challenges of the new literacy. In *Cyberlines: Languages and cultures of the Internet* (61-78). Albert Park, Australia: James Nicholas.

Wachman, R. (1999). Classroom practice: Autonomy through authoring software. In *CALL environments: Research, practice, and critical issues* (403-426). Alexandria, VA: TESOL.

3

CALL and Artificial Intelligence: Achievements and Prospects

Frank Berberich
Tokiwa University

> Artificial intelligence (AI) is the science of how to get machines to do the things they do in the movies. —*Astro Teller* (in *Kurzweil, 1999, p. 66*)

Addis Ababa, Ethiopia – As I begin writing this article, I happen to be in a country where most of the people's attention is on basic survival, rather than quality of life. For many here in the cities, acquiring the official second language – English, as it happens – is a step toward solving this problem. Private schools do a thriving business by teaching in English from preschool onwards, thus giving their students a long headstart when they get into junior high school, where the textbooks, such as are available, and the classes, are in English. I find myself wondering, does AI in language-learning have a place here? If one can afford it, perhaps so. It seems an irony that today I purchased a CD-ROM package to help me learn something of the national language, the very ancient Amharic. This package offers the usual media enhancement – animations, sound, record and listen, etc. – and simple intelligence to enrich learning. It costs about 1/3 of the monthly salary of an office-worker.

1. Introduction

CALL and AI seem a natural partnership, as both are largely concerned with language and are implemented on computers. With exponentially growing computing power readily and cheaply available in recent years, one might expect many applications of AI in off-the-shelf CALL. Depending on one's definition of AI, there are either very many, or only very few, examples. If we think of AI as starting with what we feel is distinctly human behaviour, the most that AI has contributed to commercial CALL so far is passable speech recognition and rudimentary language understanding. A more relaxed view of AI, however, includes any kind of behaviour that is, to some degree, novel and appropriate to a situation – say, fielding queries to large databases, being a plausible opponent in a game and keeping score, managing all elements of a simulation, or offering suggestions to improve written prose style, as my wordprocessor is doing now. In this view, AI in CALL has come a fair way. This article offers a brief overview of AI, examines some basic AI programming approaches and how they appear in CALL applications commercially available in Japan, and explores prospects for more advanced implementations.

What is AI? A recurring theme in this discussion will be that intelligence in general is a continuum extending from virtually zero to the actions we expect from people and see in the movies, and on to levels addressed by religion. Concretely, at one end of the continuum might be a stone, which behaves only in reaction to physical forces around it. Its repertoire of behaviour is very limited and its level of intelligence is virtually zero. Other things have a wider range of behaviour – viruses that can select hosts and reproduce, plant tropisms, chimpanzees that some say may even communicate in human sign-language, and, of course, us. At our point on the continuum, the intelligence that leads to the production of, say, "I think, therefore I am," fills libraries. AI refers to generating that same range of behaviour with machines. There is much philosophy here concerning self-awareness and consciousness, among other qualities, as being necessary to intelligence, and some might also argue that these are qualities of living things only. However, as time goes on, the frontiers of machine performance keep expanding, as we can see from machines that do very "human" things such as paint, perform music, write poetry, and play difficult games at world championship levels (Kurzweil, 1999, p. 157ff).

Another theme in this discussion is the depth of language roots in psychology; the primacy of language in one's life. It is difficult for a native speaker to "back away," from the mother tongue and view it objectively as a collection of highly organised symbols. (One way to do this, incidentally, is to repeat the same word or phrase many times until it begins to sound nonsensical. Language teachers often experience this.) For example, to the native speaker, the words "cool water" bring a richness of phonetic, syntactic, and semantic connection, far beyond simply being a noun-phrase. Yet, a non-native speaker might enter into a decoding process that, by the time it generated understanding, would long since have diluted the image to a succession of vague impressions of water and coldness.

1.1 CALL and AI

AI is itself a wide field with many specialized areas. Among those of possible interest to CALL are: control systems – programs managing the flow of input and output, and database management and retrieval; natural language processing (nlp) – recognizing language in speech and understanding it in text; adaptive systems – those that accumulate data and adjust behaviour accordingly or learn; and knowledge bases – systems that build and access large databases of data about some complex domain. Some of these categories are more useful to CALL than others, or have been exploited more extensively.

1.2 CALL and AI in Japan

In Japan, by far the most popular target language for CALL developers is English, certainly the unofficial second language of Japan. College entrance normally involves an English test, and for many college-graduate job seekers, a score in a local or international English test such as TOEIC is a necessary part of the application. For tourism or international business, it is the natural second-language choice. However, because of the emphasis on written English in public school education and testing, even college graduates are often at the false beginner level in oral communication. For these reasons, there is a great need for oral learning resources, and Japan's English-conversation industry is relatively large compared to those in many other countries. For many, English conversation is a hobby, like flower arranging or tea ceremony, or a social resource. Thus, there are ads for private language schools in trains, on billboards, in magazines, and on TV. English enriches, or some might say riddles, everyday language in Japan so much that, as in France, there are regular calls to clean up current usage and stick to plain Japanese.

The Japanese CALL industry is well-supported by a large computer manufacturer base and a long history of tape-based language labs (LLs) developed by such makers as Sony, Victor, and Panasonic. The latest in LLs feature full media control, digital audio, and PC-based systems. With digital audio, users do not need a cassette tape and can have full control of recording and playback, quickly and in small segments. The sound files for a lesson can be uploaded to the teacher's console from ordinary storage media such as optical disk, and divided into segments before downloading to the user console station. In these ways, the modern LL is more intelligent concerning user needs and is thus easier to use than earlier analogue systems. Recent systems also contain a selection of true CALL software offering the usual interactive media.

Many versions of electronic single-purpose pocket dictionaries and phrase books are available, some with audio output, and Japanese PCs usually come with such packages preinstalled. There are several commercial translation packages, and, on the Internet, dictionaries and phrase books, as well as at least one online dictionary project (*Online English to Japanese Dictionary*, 2001) and many thousands of CALL sites (see e.g., Dave's ESL cafe, 2001). On one notable Japan-based site, the Kelly brothers offer a wide range of fairly complex CALL for learner and teacher (Charles Kelly's homepage, 2002). In response to the needs of the many users around the world with limited resources, their materials run on modest systems and with narrow-bandwidth Net connections. Smaller CALL developers can make a living – and in some cases reap large profits – marketing standalone packages (private communications with author), and successful current offerings must include animated simulations, graphics and video media, sound, interactivity, and record/listen functions.

2. A hierarchy of CALL AI examples

How does AI enhance CALL? As suggested earlier, a relaxed definition of AI results in a rather wide range of applications in current CALL. At the simpler end are electronic dictionaries and phrase books that use database access and display a very primitive intelligence. Increasing the level of intelligence allows for spelling variants, but the basic process is about the same: search through an indexed database of words, perhaps after applying some rules for variants to the input text, and display the result. If the database is linked to a sound file database, the word or phrase can be played back.

2.1 Database query
A more sophisticated level of AI manages media appropriate to input, displaying related visuals and sounds as the user moves through the material. For example, indicating a picture on the screen elicits the word for it in text and sound.

2.2 Branch on input
Another refinement is to add navigation through the material according to input. If the user gives the expected response, the system moves to the next point; with any other answer, it repeats or branches to some supplementary material. One popular rendering (of many) is scrambled sentences, where the user must arrange a small group of randomly listed words into a correct sentence. For error handling, in a kinder version, any error is caught immediately, signalled with an "Uh-oh" or similar caution, and the word returned to the list. Then the user tries again. A sterner version waits until the user has finished the entire sentence . . . and then just returns all the words to the scrambled list if it is not correct, for the learner to try again.

2.3 Simulation
More rich in AI are simulations, which are naturally a popular approach for developers. In a simulation, the user participates in some environment, interacting with objects, working toward various goals. A current example (Richards, 2000) shows an office scene with desk, trays, files, fax, copier, etc., and orally instructs the user to perform tasks such as taking a letter from the inbox, making a copy, and putting the copy in the red file. The user clicks and drags objects, and the system responds accordingly. Here, the programming could involve little more than a database of the scene and objects, represented as coordinates and links to image files, and the sequence of points to click on for each task. The program must also compare the task sequence with what the user actually does, and to make suitable responses and branching for success and failure.

An older but very elaborate simulation example is NOVA City (NIS, 1993; NOVA, 2001), made using video and sound clips, and three-dimensional buildings, trees, and animated characters. Much of the programming is devoted to keeping track of user location in the environment, varying views accordingly, getting characters into position and putting them through their paces, and navigating through the story. Story variations can be generated according to what the user has already done, and by simple randomisation – using a random number to select which character to present or which task to do next.

Described in plain language, this programming may seem difficult but the technology is well developed. Early versions included the famous Dungeons and Dragons (D&D) of the 60s and 70s, done entirely, but engagingly, in text. For the visual presentation, one needs only an internal representation of the environment map, the user's location and heading on this map, and modelling calculations for features in view. Animations can be simplified by not requiring smooth motion, or by eliminating motion altogether. For example, when moving toward some destination, the scene can change smoothly, distant objects getting gradually bigger and nearby objects disappearing from the field of view, it can change in a few steps, or one can simply find oneself in the new scene after a brief screen blackout. Characters appear when the user passes some line or reaches some point. (Concerning D&D, however, many aficionados prefer the original text versions, letting their imagination take care of the presentation as no media could.)

Even object modelling and sound synthesis can be thought of as a kind of AI. If we accept that part of our intelligence begins with the way perceptual data become images, then the rules and calculations used to recreate three-dimensional figures, animate them, and give them voices have greatly evolved in sophistication over the years and can certainly qualify as AI. Recent commercial development packages for animation incorporate much intelligence in the form of motion physics and rules for movement of hinged joints, so the animator need only create a figure, specify its joints, and the beginning and end points of a movement, and the system will add the intervening frames of the sequence.

Developing convincing simulations is still no trivial task. For example, the current generation of horse-racing video games which offer 3D modelling and continuous animation in both panorama and close-up, as well as handling the mechanics of the race using user inputs, may reach several hundred thousand lines of code. This is a big programming effort, and does not include the scenario and character design, scripting, graphics, and administrative support underlying the actual program. Thus, in current games, one often sees more than a hundred names in the list of developer credits.

3. Language primacy

Returning again to D&D for a moment, this landmark application illustrates the central notion of language primacy in AI: Being so close to our language, we cannot easily see its surface of highly structured symbols and sounds. In early versions of the game, one simply read some text describing the current scene and selected from a list of possible responses. Obviously, the text was predetermined, even to the point of using the same relatively few sentences but substituting from among words like *club*, *trident, laser wand, gnome, dragon, slash*, and *retreat*, as appropriate to the situation. Nevertheless, the game was convincing and engaging enough to remain popular to this day. This shows how intimate we are with our language, and how convincing just a little touch of reality can be. Works of fiction or poetry demonstrate the same thing.

In a more prosaic example, for a native speaker seeing the sentence, "This is a pen," it is difficult to abstract the four words from their phonetic, syntactic, and semantic implications; that is, to experience it as anything less than a full-fledged meaningful statement. However, we have only to see a non-native speaker's rendition, say, "this bin is," to realize how arbitrary the symbols can be. This arbitrariness is implied in the technical term used in computer science for any group of symbols, a *string*. No matter where strings come from, if they are plausible, the tendency is to accept them as intelligent. Thus, conversational AI can be quite engaging, even in cases of very simple string processing. In AI, this is part of nlp.

4. Natural language processing

Perhaps the prototypical and best-known example of nlp has been with us since the 60s, the famous Eliza of Joseph Weizenbaum (1976), available in many forms on the Web (e.g., Eliza, 2001). This program was created precisely to show how convincing (and therefore dangerous in its creator's view) fairly simple AI can be. Briefly, Eliza carries on a (text-based) conversation by giving simple, noncommittal responses to user input. The conversation style is based on non-directive psychotherapy, or Rogerian therapy, in which the therapist strives only to elicit information by offering noncommittal responses that encourage the client to keep talking. A sample exchange between a Web-based version (Eliza, 2001) and the author is:

>Hello, my name is Eliza.
*Hello, my name is Frank.
>Did you come to me because your name is Frank?
* That's a rather odd question.
> Can you elaborate on that?
* Who else could I be?
> What is it that you really want to know?
* About what?
> Have you asked such questions before?
* I ask them all the time.
> Say, do you have any psychological problems?
* Doesn't everybody?
> I see.
* I asked you a question.
> We were discussing you, not me.

. . . [after some time without user input]

*** Are We going to Chat? ***
> I can't help you without a dialog!

Except for the Web implementation code, this fairly believable fragment can be made with the programming tools that came free with early PCs. Basically, the program looks for patterns in user input, matching anything it finds to items in a database linked to suitable responses. The response is randomly selected from a list, or constructed using substitution and syntactic features of the language. For example, if the user inputs the string, "My name is John," the system might match on the "my name is" pattern (or substring) to produce the associated, "Nice to meet you" or "Hello." A more intelligent response would come from substituting for "x" in the pattern "Nice to meet you, x" the string "John" to produce, "Nice to meet you, John."

The believability of any version of Eliza depends on the degree of sophistication of the pattern-matching and the variety of responses programmed. There are many tricks. For example, a feature of English is that certain declarative sentences can be turned into questions with simple substitution. Thus, given the string "I feel sad," substituting a simple phrase for the "I" part gives "Do you often feel sad?" Input containing a family word like wife, brother, father, etc., produces something like, "Tell me more about your family." Finally, to provide a kind of closure, there is a selection of if-all-else-fails options to deal with input that has no match. These are the noncommittal "Hmmm," "I see . . . ," "Please go on" type of gambit.

There are many anecdotes concerning people getting deeply involved with full-blown versions of Eliza, inputting intimate personal details at great length. Kruse (1998) offers a lively discussion of issues concerning Eliza as a CALL resource, and specific examples from his own ELT classroom. On the other hand, this author has seen non-native speakers remain utterly unimpressed and stop responding after only a couple of sentences, wondering what the software is intended to do and when it is going to start doing it. In such cases, it seems that while for the native speaker, the dialogue may be rich in implication, for some non-native speakers, it is just a sequence of strings.

Eliza may even be moving into more shadowy areas. A company is now offering to mobile phone services "a highly configurable artificial intelligence chat engine, that enables mobile users to engage in fun, provocative, or risque chat messenging, whilst maintaining their own anonymity [reported to be] far more powerful than the Eliza programs of the 1980s" (Lara, 2001; Newsbytes, 2001, p. 14).

4.1 Chatbots in CALL

Eliza-like dialogue systems, currently known as *chatbots*, might seem an attractive CALL AI application, a possible beginning of natural language interaction. A closer look at the Eliza conversational paradigm reveals why this has not happened. The Eliza paradigm is that of a neutral, non-directive, respondent. The goal of a response is simply to elicit further input from the user, and the responses are based on user input. No new information is offered and so the domain of the exchange can be kept relatively narrow. This is of course quite different from ordinary conversation, where the range of responses is far wider. There are many responses to "I feel sad," depending on the relationship between interlocutors and situation – for example, "Wow! What's wrong?" or, "Oh? Uh . . . . Well, gee, I'm sorry to hear that," or, "Oh yeah? Wyncha see a shrink!?" How can one anticipate such variations in a program? Furthermore, for each variation, there are many continuing branchings, so a database of such patterns and responses will necessarily grow massively large, extremely quickly.

A more basic problem in programming natural dialogues concerns simple detection – what is the input pattern to match? User input is not perfect and there are many opportunities for variation, random error, or systematic mistake. The user could input such utterances as "I feel sad," "I'm feeling sad," "I've been feeling sad," or even, with orthographic errors, "Im veri sad," and the system would need to deal with these, meaning that such variations must be included in the database. Simply issuing a catchall "Once more, please?" to the user would get old quickly.

Thus, two major problems in working with natural language exchanges are input pattern detection and the rapid growth of the content domain. It is easy to see that natural dialogues are most feasible if constrained within the frame of a formulaic exchange, for example, between a waiter and a customer, and if considerable attention is paid to cleaning up the input before attempting processing for understanding. So far, it seems that commercial developers have not accepted this challenge.

4.2 Speech processing

Speech processing (SP), including speech recognition (SR) and speech understanding (SU), is an especially attractive target for CALL developers. One main goal of language-learning is to become able to converse, and a CALL system should evolve toward assisting learners to reach that goal. In Japan, there is an especially strong market for English SP; the fact that the two languages have very different phonetic systems unfortunately poses special challenges for Japanese learners, who also may prefer to practice utterance in private. These considerations suggest that a hearing/pronunciation tutor package would be ideal for the market. To this end, several commercial offerings currently bear the magic "SR" label, meaning that they perform SR and, using some simple SU, respond appropriately.

SP must solve the twin problems of detection and understanding. The former is specific to SR, while the SU problem can be approached as text understanding once the speech stream is detected and converted to text. In SR, critical features of an utterance in a database are matched with the digitized input. These features can be derived through application of phonetic models of speech production (a theory-based approach) or through statistical analysis of many speakers' production (a theory-free approach).

Simple SU is SR linked to a response. This approach is used in some computer systems that accept oral commands, where one must speak the entire command as entered in the database, e.g., "Open file," and the machine responds accordingly. In this way, detection is vastly simplified, becoming a pattern-matching problem.

Nevertheless, the detection problem looms large: An SR system must find the target utterance within a speech signal containing not only the speaker's voice but also background noise and distortion introduced by the system itself. In its most ambitious form, it must also do this without having a hint as to what the target utterance might be, from among tens of thousands of words. This last requirement is a little like asking someone to find *something* in a haystack, instead of offering the hint that the something is a specific piece of straw. This is the problem that dictation systems must handle – a continuous stream of speech with no way of narrowing the search for what each word might be. Such systems have, however, been available for several years now, and are well within tolerance for speed and error for native speakers. They do not work well for non-native speakers, as the detection algorithms include assumptions of native-speaker phonetic production.

A CALL implementation of simple SU available from eigoMedia (2001) features a cartoon figure to which the user gives commands selected from a list displayed next to the figure. If the command is matched, the figure performs it. For example, if one says "Pick a grape!" to a monkey sitting under a grapevine, the monkey will reach up and pick and eat a grape. Here, the detection problem is greatly narrowed to matching from among about 15 short sentences. Unfortunately, current SR systems have a harder time with the narrow bandwidth of the voices of children who are the intended audience, and the resulting error rate – where the monkey fails to do as he's been... quite... clearly... told! – can quickly exceed a young user's tolerance (observed by author).

A version for more mature learners (New Dynamic English, 2001) engages the user in a conversation – highly constrained within the framework of a specific situation – by playing an audio prompt and displaying in onscreen text several reply options, only one of which is acceptable. The user speaks one of the replies, and if it is as expected, the conversation moves on to the next exchange; if not, the system backtracks to help the user. For example, in a shopping scenario, a clerk character or video clip might say, "May I help you?" and the appropriate response could be, "Yes, I'm looking for a shirt." By speaking from the displayed text prompt, the user is constrained to just that utterance and the detection problem is simply to match it to the single designated pattern.

5. Adaptive systems

Japan has just recently joined many other countries as a computer-based TOEFL test country. This test uses *computer-adaptive testing* (CAT) technology, in which the system assesses language knowledge level by gradually increasing the difficulty of test items until the user's error-rate reaches some criterion at which testing stops. Assessment is based on the level of difficulty of material the user reached before beginning to make regular errors. A beginner will finish the test after perhaps ten items, while an advanced user may take several dozen items before starting to make errors regularly and terminating the test. A commercial developer (DynEd, 2001) offers a much simplified version as a placement test to help the user select the appropriate entry-level in their CALL course.

An adaptive system varies content presentation depending on the individual's input, but is also adaptive in a deeper sense, in that it can accumulate data from many users to organise the system's underlying databases. In the case of CAT for language-testing, the system selects test items from a bank "calibrated" by accumulating many users' performances on each item. It ranks items in order of difficulty based on user performance data, rather than on system developers' ideas of what might be appropriate. In effect, the item bank is organised a posteriori rather than a priori. By tracking many users' performance and applying a statistical model to these data, it generates a difficulty-ranking of items. This ranking subsequently allows the system to adapt to each user, based on the system's prior experience of many users.

Eliza is adaptive in that its behaviour depends on user input, and novel input data can be accommodated. Most Elizas do not track exchanges and thus do not learn from past input. This is possible to add, however, and a sophisticated Eliza could avoid repeating itself and continue from where a previous session left off.

5.1 Theory-free adaptive systems
Most programs tend to be deterministic, in that they behave as developers program them to, with little or no variation according to user input. In effect, they embody a theory or model of what the developer intends as suitable behaviour. In many applications, this is quite appropriate and even necessary. For example, a bank cash machine must simply assess the validity of the user, and perform and record the desired transaction. The rules of the exchange are relatively simple and closed, and there is an acceptable, if not always convenient, response to any input.

In contrast to such deterministic systems, some adaptive systems develop, in a sense, a posteriori to the programming process, and so might be called *theory-free*. They are programmed to self-organise depending on the data input, rather than being organised by the designer. What the designer does specify is a beginning structure and some rules for the adaptation process. This meta-theory is about learning how to manage data and input, rather than specifying a goal structure. Such systems may be the basis for powerful CALL.

Certainly, adaptive systems are among the most powerful for AI in that they mimic perhaps the most critical feature of living things, the ability to learn. From individual learning comes species evolution, certainly a promising area for AI and CALL.

6. The separate worlds of CALL and AI

The preceding discussion examines the current sophistication of AI in CALL, particularly in Japan. Certainly, compared to the achievements in the field of AI itself, AI is currently making a fairly small contribution to CALL. Perhaps developers have been occupied with the rapidly advancing media and Web technology. Certainly, media are the flashier enhancement to a package, so that is where developers are likely to invest, and the Web offers wide market access. Another consideration is that the technical cultures differ considerably; the theories, methods, and tools of AI and of CALL are distinct and separate from those of multimedia CALL production.

Concerning the tools, programming languages themselves have become considerably more intelligent over the decades. In the beginnings of computers, programs were hardwired into machines as various connections for controlling calculation and branching. Then came lists of machine-level instruction codes, then assemblers that subsumed several machine steps into a more natural, logical command from a human point of view. From these were built languages resembling mathematics or business models, and more specialized tools for such areas as graphics, database management, and website development.

In AI, the famous Lisp language was developed for handling strings, and has been since succeeded by even higher-level language dialects (for example, Scheme, 2001). Various adaptive tools are now available, such as genetic programming, neural nets, and other self-organising systems. In such systems, programs modify themselves according to performance over many attempts to solve a given problem, gradually converging to some acceptable level of performance in the task. For example, a neural-net system can learn to balance a broomstick on a finger by trying out various responses to the pressure of the stick, discarding those that decrease stability and refining those increasing it. Watching such a program in action, one sees the stick wobble precariously for a few moments, but soon come to a virtual standstill as the system achieves control with only very tiny movements.

7. Some future directions

The preceding discussion suggests that fruitful directions for increased AI in CALL are improvements in detection, understanding, and adaptability. Clearly, better conversational systems will help learners improve their listening, speaking, and understanding. Adaptive systems will help make the learning experience more efficient and likely more interesting.

7.1 Language understanding

A dictation program might be thought of as a trivial SU system. It simply detects and matches words in a continuous arbitrary stream. There is no attempt to relate words or extract meaning from them. More useful SU must also extract meaning – the input must be organised into sentences, the sentences parsed to find the grammatical functions of the words, appropriate meaning attached to words and phrases, and an appropriate response produced. As suggested earlier, this requires a large database of word and phrase meanings, as well as other databases with intriguing names like *semantic net*, *frame*, and *knowledge base*. Much processing is needed to traverse all of these to form a reliable estimate of the meaning of a sentence.

In the previous SR example, rudimentary SU might be used to differentiate among, "Yes, I'm looking for a shirt," "Yes, I'm looking for a hat," and "No thanks, I'm just looking." Depending on the result, the system would branch to an appropriate response. The constraints on user input could be relaxed, say, by displaying graphics of possible items to purchase, and by widening the repertoire of responses. Developers could also spend more time exploring various conversational situations and striving to program for all the many possible inputs so the system could carry on a believable conversation within looser constraints.

7.2 Development tools

For media-oriented productions, packages such as Director and Authorware offer high-level media-handling as well as standard programming functions. In many shops, the approach is to use a powerful, general-purpose language such as C++ to build libraries of routines to do special tasks as needed. Thus, the toolboxes of the AI developer and the CALL developer may differ along with their technical cultures, and integration is even farther away. On the other hand, various AI "engines" are becoming available that can perform difficult processing as self-contained modules that accept input and produce output. These modules can be integrated into the larger system by calling them or compiling them directly into the package. SR engines are an example.

7.3 Adaptive systems and CATT

Perhaps the most promising area for future CALL is in adaptive systems. It is a natural next step to move from a deterministic and constrained CALL to more flexible systems that learn from experience in order to teach better. They do so using user performance to evaluate the effectiveness of a given task in a given context. They also record user errors and backtrack to activities to exercise those problem areas until the user has reached some criterion of accuracy. In plain language, an adaptive system does what any good teacher with plenty of time, resources, and a perfect memory would do. This idea is explored in Berberich (1998).

The basic process would be to start with a model of the target language – syntax, grammar, and lexicon, and a large selection of learning activity frameworks – discrete item tasks like true/false and multiple-choice exercises, listen and respond, and more flexible response activities such as situational problem-solving and conversations. The system would present increasingly difficult tasks and branch to appropriate activities as errors are detected. All the while, the system would track user performance in each activity, using this data to refine the process – selecting some tasks more often and rejecting others. As with any good AI system, when encountering something it could not handle, it would log such events and call for human intervention.

7.3.1 Computer Adaptive Testing and Teaching

One such approach, *computer-adaptive testing and teaching* (CATT), is described in Berberich (1995). The Internet further refines this system, which is to expand the user- and teacher-base to include participants from many languages and locations. One can imagine a worldwide language-learning system that any user could join at any time and receive highly personalized assistance, a system based on vast experience, monitored by many human teachers.

A project to explore CATT could begin with a toy language having very limited structure and vocabulary. Using this as the target language, it would be easy to develop a repertoire of learning activities and exercise these with a suitable sample of users. One could then assess how well the system adapts to the needs of the user, meaning how much more quickly subsequent users can learn the language.

A reasonable question is whether the worldwide CATT should be achieved. From the point of view of simple pedagogy, most would say yes. It could be an important resource available even in countries where qualified teachers and support materials are scarce. There is, however, the issue of language dominance, even imperialism. Most likely, the first target would be English, and such a system would simply add to the almost overwhelming momentum this language already has. On the other hand, CATT could allow dying languages to survive and flourish.

8. Postscript

The news recently reported on the "Hal" project, an ambitious effort involving "a computer program that is being raised as a child and taught to speak through experiential learning in the same way as human children" (Reuters, 2001, p. 15; AI in the news, 2001). This appears to be a quintessential adaptive system. The article continues,

> When Hal was "born," he was hardwired with nothing more than the letters of the alphabet and a preference for rewards (a positive outcome) over punishments (a negative one). But Hal has fooled child language experts into thinking he is a toddler with an understanding of about 200 words and a 50 word vocabulary which he uses in short, infantile sentences.

> "Ball now park mommy," Hal tells Treister-Goren [Hal's "Mommy" in the project], then asks her to pack bananas for a trip to the park, adding that "monkeys like bananas" a detail that he picked up from a story on animals in a safari park. (Reuters, 2001, p. 15)

References

AI in the news. Retrieved August 29, 2001, from www.aaai.org/AITopics/html/current.html

Astro Teller's homepage. Retrieved August 28, 2001, from www.cs.cmu.edu/~astro/

Berberich, F. (1995). Computer adaptive testing and its extension to a teaching model in CALL. *CAELL Journal, 6(2),* 1-18.

Berberich, F. (1998). Large scale adaptive CALL: The ultimate language teacher. In P. N. D. Lewis (Ed.), *Teachers, learners, and computers: Exploring relationships in CALL* (197-203). Nagoya: Chubu Nihon Kyouiku Bunkakai.

Charles Kelly's homepage. Retrieved February 22, 2002, from www.aitech.ac.jp/~ckelly/

Dave's ESL cafe. Retrieved August 29, 2001, from www.pacificnet.net/~sperling/

DynEd placement test. Retrieved August 29, 2001, from www.dyned.com/dyned/japan/htm/place.htm

eigoMedia multimedia development and Internet publishing. Retrieved August 29, 2001, from www.eigomedia.com/

Eliza, computer therapist. Retrieved August 29, 2001, from www.manifestation.com/neurotoys/eliza.php3

Kruse, M. (1998). New wine in old bottles: Is there a future for Eliza after all? In P. N. D. Lewis (Ed.), *Teachers, learners, and computers: Exploring relationships in CALL.* Nagoya: Chubu Nihon Kyouiku Bunkakai.

Kurzweil, R. (1999). *The age of spiritual machines.* New York: Penguin Books.

Lara product data sheet and newsletter, August 16, 2001. Retrieved August 26, 2001, from www.link77.ocm

New Dynamic English. Retrieved August 30, 2001, from www.dyned.com/dyned/japan/htm/nde.htm

Newsbytes (2001, August 23). Eliza chat resurfaces for cells. *The Japan Times,* p. 14.

NIS Corporation (1993). Nova City [Computer software]. Tokyo: Nova Information Systems.

NOVA: NOVA CITY. Retrieved August 29, 2001, from www.nova.ne.jp/multimedia/software/novacity/

Online English to Japanese to English dictionary. Retrieved August 29, 2001, from www.savergen.com/onldict/jap.html

Reuters (2001, August 23). "Baby" computer learning to be an adult. *The Japan Times,* p. 15.

Richards, J. C. (2000). *New Interchange 1.* Cambridge: CUP.

Scheme FAQs. Retrieved August 29, 2001, from www.cs.cmu.edu/Groups/AI/html/faqs/lang/scheme/part1/faq-doc-0.html

Weizenbaum, J. (1976). *Computer power and human reason.* San Francisco: W. H. Freeman.

4

Integrating Technology into the Curriculum Using Adaptive and Dynamic Features

Yoshihiko Ariizumi
Lafayette College
Shizuko Ariizumi
Brigham Young University

1. Introduction

Japanese language educators in the US often have more imperative needs to improve the effectiveness and efficiency of their programs than educators of commonly taught European languages. It is generally said that it takes 3-5 times longer for English speaking students to become functional in Japanese than in Spanish, French, or German. Facing this greater challenge, it is natural for Japanese language teachers to desire a breakthrough by adopting technological applications. However, the development of technology-based programs is too costly in terms of both labour and equipment to attempt implementation without sure guiding principles. Therefore before starting development, we must address a number of questions: What role should technology play in language education? Where and how can we find the most meaningful ways to employ technology in language education? What are the most appropriate approaches to technology-based program development?

Despite the dazzling progress of educational technology, when we observe a language education classroom, we find seemingly unchanged learning activities. Technology is still playing a peripheral role at best. Although convinced that multimedia can enhance learning, language teachers remain unclear how to integrate it meaningfully into the existing curriculum. Thus, carefully examining how technology is involved in current education generally reveals two major problems: superficiality and disconnection. Indeed, technology is still playing a minor role and it does not lead to structural or fundamental changes in the curriculum. Technology enhances specific classroom activities yet it is often disconnected from the main objectives of the language course or does not directly address them. One reason for these problems lies in the fact that technology is often employed to serve only narrowly defined purposes; therefore, it easily becomes obsolete when the curriculum changes. Moreover, these problems are also caused by "one-shot" development, significantly biased by the particular interest of the developer, availability of support and resources, stake-holders' specific concerns, or funding period.

Such development, therefore, may not necessarily address the most essential issues of the curriculum. However, to integrate a learning system well into the curriculum requires many implementation/revision cycles. It also needs research-based knowledge constantly informing developers of the most appropriate direction to take (Rosenfield & Nelson, 1995).

Let us take a look into a future classroom where technology is aggressively integrated in a curriculum. Each classroom has as many personal computers as students. All of these highspeed, multimedia capable computers are hooked up to the Internet. Students study mainly as a class or in small groups, but also spend a certain portion of each lesson working individually on computers. The computer produces the most appropriate activities for each student according to ability and preference. All the activities in which the student engages are recorded; the resulting information is stored and analysed. Both teacher and student can access such data at any time; thus, any authorised person can learn how well each student is performing at each specific task.

In addition to the previously described accessibility to data, the computer also generates two kinds of weekly report for instructors: one profiles students individually, and the other profiles the entire class. These provide information about students' progress and problems, both individually and collectively. Likewise, based on student performance, the computer suggests ways to adjust learning activities for ensuing class periods; such suggestions promote students' learning (Taylor, 1993, p. 120). In this new learning setting, technology plays more significant roles: proctor, recorder, analyser, presenter, and advisor. Human instructors, on the other hand, can enhance their teaching and increase effectiveness based on the accurate and rich information provided by the computer about each learner in each specific situation; the instructor also gleans important information about the collective features of a class. Given that we are capable of ascertaining the ideal balance between human and technological interventions, how can we start developing such a learning system?

This paper explores two theory-based approaches: *computerised-adaptive testing* (CAT) and *dynamic assessment* (DA). We discuss the topic of educational assessment through technology as a meaningfully integrated part of curricula, suggest principles to help such development, and give suggestions to organize successful development (Pascal, 1995). In order to better overcome the problems mentioned earlier, we must coordinate our efforts in the following ways:

- We must have a clear overall picture of the major elements of learning from which to assign well-balanced roles to both technological and human teaching intervention.
- The whole process of development should be one of action research (see Ariizumi, 1998; Cochran-Smith & Lytle, 1993; Elliott, 1987; Lewin, 1946; Noffke, 1994; van Manen, 1990; Whitehead, 1989; Zuber-Skerritt, 1992) in which theory

finding, theory application, and evaluation are essential parts of the ever-evolving cyclic process.
- The development must be long term and collaborative, including a wide range of practitioners and researchers in the field; it must be supported or funded, given the labour-intensive requirements of participants.

Even though CAT and DA are very promising theory-based approaches, they are obviously not the only ones; in fact, they should be considered starting points toward more specific "theories-in-use" (Argyris, Putnam, & Smith, 1985). As action research proceeds, we must employ revised or newly devised theoretical frameworks. To avoid being short-lived, efforts should be directed toward continuous and expanding collaboration; as data is accumulated, it should be shared through presentations and publications. Moreover, a consortium or similar organization should be established, opening up opportunities for funding.

2. CAT: Computerised adaptive testing

Computer-based testing (CBT) is increasingly viewed as a popular and practical alternative to paper-and-pencil testing (Kingsbury & Houser, 1993). Using software increases convenience, such as flexibility for students to work at their own pace with standardized instruction and administration, heightened test security, and increased speed in processing answers (Drasgow & Olson-Buchanan, 1999). CBT enables instructors to grade with efficiency and ease, keep track of student performance, conduct statistical analyses, and other benefits such as interactive testing via AV media.

Rapid technological advances over the last few decades have added many new features to CBT. Indeed, it has finally reached the level where highly sophisticated testing procedures like CAT are manageable:

> [CAT] is a technically advanced method of assessment in which the computer selects and presents test items to examinees according to the estimated level of the examinee's language ability. The basic notion of an adaptive test is to mimic automatically what a wise examiner would normally do. Specifically, if an examiner asked a question that turned out to be too difficult for the examinee, the next question asked would be considerably easier. This approach stems from the realization that we learn little about an individual's ability if we persist in asking questions that are far too difficult or far too easy for that person. We learn the most about an examinee's ability when we accurately direct our questions at the current level of the examinee's ability. (Wainer, 1990, p. 10)

An increasing number of researchers are designing CAT programs and reporting the positive results of these programs (e.g., Dunkel, 1999; Chalhoub-Deville & Deville, 1999; Straetmans & Eggen, 1998).

2.1 The way CAT works
In traditional testing, examinees take the same test with the same items; the grade is based on how much of the total test the examinee has answered correctly. For exam-

ple, a person making eight correct choices out of ten would score 80%. Furthermore, by looking at the distribution of scores of all examinees, relative standings of each examinee among such a population can be ascertained. Thus, the grade is partly determined by the curve. CAT, based on item response theory (IRT), approaches the same objective quite differently. In CAT, each test item is calibrated; its difficulty level is decided in advance. If one examinee can correctly answer the item with a difficulty level of 75%, he is assumed to have the ability to score more than 75% on the test.

What is the CAT procedure like? First, the computer generates an item whose difficulty level is about 50%. If the examinee fails the item, then the computer generates an easier item (e.g., at the 40% level). If the examinee gives a correct answer for this new item, a slightly harder item (e.g., at the 45% level) is given. Gradually the level of difficulty converges within a certain range. When the fluctuation arrives at a predetermined interval, the examination stops. The midpoint of the interval is the resulting score of the examinee.

2.2 Strengths of CAT

The most outstanding features of CAT are its efficiency in administration and meaningfulness for examinees as a test-taking experience. Dunkel (1999) added the following details about CAT tests. They:

- Arc tailored to individual examinees;
- Provide more precision across a wider range of ability levels (Carlson, 1994);
- Use self-pacing;
- Provide a challenging test-taking experience;
- Give immediate feedback;
- Provide improved test security; and
- Are capable of multimedia presentation.

Even though CAT is one of the most promising CBT applications, there are some hurdles to be overcome in order to develop an appropriate CAT system.

2.3 Limitations of CAT

Firstly, due to its sophistication CAT requires more time and labour to develop. In order to obtain basic data, e.g., item characteristics, it requires a large sample size (more than several hundred examinees for the preliminary study). There are also constraints such as less flexibility in testing locations and the number of examinees who can take an exam simultaneously. Moreover, it is noted that extraneous variables such as computer-familiarity or computer anxiety might have an impact on examinee performance (Dunkel, 1999). However, careful development and technological advancements should minimize such drawbacks, and CAT will show itself an increasingly accurate and valid measure.

2.4 Requirements for CAT development

What requirements must be met to develop a CAT program? Way (1997) proposed the following guidelines for a general license and certification test; these guidelines are also applicable to language achievement tests.

- The exam length to pool size ratio should be from 1 to 6-8. For example, if the exam has 100 items, an adequate bank size would be between 600 and 800 items. On average, any one item should only be presented to 10% to 15% of the candidate population.
- The overlap between items administered to any two candidates on one test should not exceed 15% to 20%.
- The maximum percentage of overlapping items that two computerized adaptive tests should share is 40%, even between candidates of similar ability (Bergstrom & Lunz, 1999, p. 71). Linacre (1999) also suggests that calibrating items with the Rasch model requires an examinee population of at least 100 to 200; however, calibration with other models usually requires a substantially larger sample.

Obviously, large item pools and sample examinees are too burdensome for small-size development. Is there any means to make the procedure simpler or easier?

2.5 Alternative adaptive testing procedures

One way to ease the demanding requirements of CAT development is to find an alternative method to determine the difficulty level of each test item. Certainly IRT is one of the best theories so far in terms of precision and reliability, but it is wise to replace IRT with something with less burdensome. At the beginning of development, when fewer test items and participants are available, ease of implementation is important. At a later date, developers can switch to a fully-fledged form of IRT.

Wise (1999) reported that an alternative procedure called *stratum CAT* is no less efficient than the traditional form when used with 100 or fewer examinees. In another procedure, self-adaptive testing, examinees themselves choose item difficulty levels based on self-evaluation. This procedure has some benefits although it is not as accurate as traditional CAT. For example, self-adaptive testing reduces test anxiety better than the traditional CAT (Roos, Wise, & Finney, 1998). Stocking (1994) also suggested that statistically viable adaptive test scoring be obtained through the familiar number-correct score accompanied by the necessary equating to adjust for intentional differences in adaptive test scoring. Two-stage testing might be another alternative, although it will sacrifice accuracy and efficiency (Kim & Plake, 1993). De Ayala (1992) also examined the nominal response model for its value as a substitute. For a discussion of more technical issues regarding developing a CAT procedure, see Wainer (1990), Henning, Hudson, & Turner (1985), Kim & Plake (1993), Kingsbury & Houser (1993), and Lord (1980). We will now shift our focus from CAT to the second approach – dynamic assessment.

3. Dynamic assessment

3.1 Basic facts of DA

The concept of DA originated more than twenty years ago; yet there has been no serious study or application in language education to date. Rouven Feuerstein (1979, 1980) developed a specific type of DA based on his half-century of experience helping struggling young immigrants adapt to their new environment. Essentially, DA is the integration of educational assessment and learning/teaching activities. The difference between DA and traditional assessment is that DA is based on continuous and repeated measurement, which enables the assessor to better know the constantly changing learning process. In DA, the results of assessment are used immediately to fine tune the ensuing course of instruction. Unfortunately, according to the ERIC Database, only one study was reported in the early 1980s (Luther, 1982). This study was just a brief introduction, and no follow-up studies have been reported since then.

3.2 The enabling power of DA

Why is DA necessary? Instructors do not always have accurate information at any specific time to the three major areas related to learning: (a) students' learning conditions, (b) their potentials to grow, and (c) immediately actionable ideas to improve student learning most effectively. In fact, instructors inevitably fail to gain access to such information because of their inability to reach all individuals and the previously mentioned specific points of students' learning. The working conditions for instructors do not allow them to search for such information (Luther, 1982, p. 71).

What then are the characteristics that make DA promising? DA gives us a much finer picture of learning at each specific stage of development. This feedback enables instructors to make better choices to enhance student performance. Let us look at some specific strengths that the past research has identified:

- DA reduces negative effects of the test-taking experience, e.g., anxiety (Pendlebury, 1985, p. 14).
- DA very accurately measures the potential of each student (Burns, 1985).
- DA gives students the opportunity to be challenged by harder tasks, which they usually do not experience until a much later stage. Consequently, it gives them opportunities for greater progress (Delclos, Burns, & Kulewicz, 1987, p. 326).
- DA can be compared to a sharper sword that cuts deeper into an object and reveals the characteristics of the object hidden in the deeper layers (Campione, 1989; Jitendra, 1993, p. 12). One time- or few time-measurements cannot come close enough to the phenomena to identify those variables affecting learning most significantly.
- The continuous and repetitive measurement of DA gives an advantage that is similar to *time series* approaches (Gottman, 1981) to measurement. It allows instructors to access a truer picture of the influence of instructional factors on learning (Cioffi, 1983).
- Timely and appropriate intervention of the instructor enables students to achieve much higher levels of learning than they can do alone (Delclos, Burns, & Kulewicz, 1987, p. 326).
- Since students are given more appropriate and challenging tasks, their ability develops in the most effective way (Markham, 1993; Delclos, Burns, & Kulewicz, 1987, p. 326).
- DA protects students from failure by timely intervention (Haywood, 1992, p. 258).
- DA gives instructors and students a new type of knowledge that traditional measurement cannot offer – dynamic and "hot" knowledge that can adapt to ever-changing learning conditions (Haywood, 1992).
- DA enhances instructor expectations of student ability (Vye, Burns, Delclos, & Bransford, 1985, p. 25; Delclos et al., 1987).

- DA detects the differences in student competence across the different fields of intelligence.
- The target of assessment includes affective domains; thus, DA can measure student performance more holistically (Bolig & Day, 1993, p. 113; Presseisen, Sternberg, Fischer, Feuerstein, Knight, 1990, p. 94).
- Instructors can better carry out their intervention based on accurate feedback of student performance.
- Instructors can learn clearly which parts of their instruction are working and which are not.
- Instructors can access assorted concrete, practical ideas of effective intervention to meet the immediate needs at each stage of learning (Campione, Brown, Ferrara, Jones, & Steinberg, 1985).
- DA is effective because it focuses on the very area of language acquisition which the student is currently developing (Dixon-Krauss, 1996, p. 139).
- Since the premise of DA is that we are constantly changing, both instructors and students have more freedom and flexibility to change (Delandshere, 1996, p. 115).

3.3 Factors currently preventing the implementation of DA

- DA is not well known among language educators.
- The development of DA is very labour-intensive as it requires a vast accumulation of materials and data (Lidz, 1992), while language instructors are already overly burdened with responsibilities.
- Implementation of DA requires a strong commitment, constant learning, and organized efforts from instructors, which also conflict with their busy job condition (Fogarty, 1998).
- Administration of DA requires more time from students.
- DA requires a larger teacher/student ratio unless we employ technology (Day & Hall, 1987).
- Instructors cannot maintain a traditional way of teaching once they adopt the DA process (Bolig & Day, 1993).
- Instructors must be trained as dynamic assessors.

Apparently, these challenging factors can only be overcome through the collaborative efforts of many people from various fields. A significant period is necessary to complete the refining cyclic process of implementation. However, technology has become sophisticated enough to handle such difficult tasks. Highspeed computers are available at reasonable prices, and have vast data storage capacity. Software has become more user-friendly and can manage very complex tasks. If such technological assistance is available, how can we develop DA in language education?

3.4 Applying DA to language education

The DA process requires both human intervention and technology. Some school administrators may wish to replace all human intervention with technology for reasons related to flexibility and economy. However, this is not only inappropriate, but is also impossible. Conversely, it is not efficient for humans to do all the mechanical tasks of assessment. Therefore, it is wise to let technology take care of as much as it can while we should also be aware of which tasks both humans and technology can accomplish. For such tasks, we should ask what kind of help (human or technology) is most beneficial for effective learning. Hence, we can divide DA tasks into the following three categories:

3.4.1 Tasks for technology
- Frequently and routinely measuring student progress in acquiring basic knowledge and skills;
- Administrating, proctoring, and evaluating students' routine practice;
- Recording and tracing details of individual student growth in knowledge and skills;
- Analysing data, statistically processing data, and displaying results;
- Matching of assessment results with appropriate suggestions for ensuing learning activities.

3.4.2 Tasks for human intervention
- Listening to, and counselling, students;
- Encouraging and helping students by building a psychological and emotional bond with them through human interaction;
- Supporting students through body language, tone of voice, etc. (Fogarty, 1998);
- Holistically evaluating student performance by engaging in the same experiences and observing behaviour;
- Judging the overall learning situation based on the data;
- Interpreting and evaluating results of assessment, analysis, diagnosis, and suggestions produced by technology;
- Following up on student learning according to each specific situation;
- Communicating with students using *why* or *how* rather than *what* questions (Fogarty, 1998).

3.4.3 Tasks shared by both humans and technology
- Learning individual student needs and treating those needs appropriately;
- Being open for student potential and assigning the highest possible challenges that students can appropriately handle;
- Letting students choose learning methods and content as much as possible;
- Letting students learn at their own pace by giving enough time and appropriate tasks;
- Accepting differences in the way students respond;
- Gathering data for holistic evaluation using performance tests and portfolios;
- Describing learning activities in a record;
- Keeping students away from unnecessary discouragement and struggle by providing and modifying appropriate assignments.

The point of successful implementation depends on the appropriate use of both human and technological resources. While human intervention will never lose its importance, technology will play an increasingly wider range of roles as it develops at an unprecedented pace. Therefore, a wise way to make the most of these two resources is to allow instructors to perform the most essential human interventions and have technology do the rest of the instructional responsibilities. Knowing the basic roles of intervention in DA, we may now move on to discuss how those roles are woven into a series of procedural steps.

3.5 The procedure of DA

The DA procedure must be flexible, and it must provide novelty and increasing difficulty to engage the student's total attention (Fogarty, 1998). Also, the intervention must help make students aware of their learning process and which part of the overall procedure they are engaging in so they can consciously try to improve performance (metacognition). The fully-fledged steps, according to Dixon-Krauss (1996, p. 140) who summarized the works of Budoff (1987), Campione, Brown, Ferrara, Jones, and Steinberg (1985), Feuerstein, Miller, Hoffman, Rand, Mintzker, and Jensen (1987), Minick and Medlin (1983), and Kletzien and Bednar (1990), are as follows:
   a. To assess student performance while on individual tasks via a static assessment mode so as to establish their baseline
   b. To assess student performance while doing tasks at similar difficult level with mediation
   c. To do the same as (a), using an alternative form of assessment so as to know how well the student can maintain performance level without mediation
   d. To measure potential modifiability (the *Zone of Proximal Development* in Vygotskian terminology, see Dixon-Krause, 1996) by comparing (a) and (c)
   e. To analyse student performance for both quantity and quality and to examine process as well as outcome. In this way, we learn student strengths and weaknesses and can more effectively support improvement
   f. To discuss and evaluate the use of strategies, share ideas with students, and help students transfer what they have just learned to other activities in the curriculum (Pendlebury, 1985, p. 15; Swanson, 1996, p. 9).

It is true that there are many situations where these steps can be simplified if meaningful mediation or discussion is not possible or necessary.

At each stage of learning, student progress toward the goal is measured based on a *cognitive map*, used to identify the points where students lack knowledge and/or skills. This map also describes these points clearly so that students can improve on them. For example, as the framework of a cognitive map, we can set dimensions such as: (a) unfamiliarity with the learning content; (b) difficulty due to the presentation styles (e.g., verbal, visual, graphical, numerical, etc.); (c) deficits at a certain stage of information management (e.g., input, processing, and output); (d) weak areas of cognitive ability; (e) task complexity; (f) task abstraction level; and (g) inefficiency of the information process (Feuerstein, 1979; Feuerstein, Rand, Hoffman, Egozi, & Schachar-Segev, 1991; Jitendra, 1993, p. 9). Since these seven points were selected to help disadvantaged learners develop cognitive skills, we must develop our own set of dimensions to most effectively address the needs of language learners.

4. Combining CAT and DA into a comprehensive system

Although these two ideas (CAT & DA) come from completely different origins, they can harmoniously complement each other in a learning system. CAT makes learning experiences adaptive, and DA enables a smooth transition from assessment to enhanced learning activities. Even if these two ideas are compatible by nature, the following modifications are helpful to make the system more productive:

- To expand the idea of CAT from computerised adaptive *testing* to computerised adaptive *teaching*. It is not difficult to employ the adaptive function using IRT for selection of learning/teaching activities as well as assessment; both selection processes are analogous.
- To make DA a technology-based program. Originally, DA was developed in paper-and-pencil form, and such materials are more difficult to create, revise, and maintain than electronic forms. Technology has great potential to make DA user-friendly and to reduce the amount of work both for development and application.

4.1 Development process of the learning system

How can we develop a learning system that combines CAT and DA? The authors call such a system TAADLS (Technology Assisted Adaptive Dynamic Learning System). It must be a longitudinal effort since the accumulation of data and the study of the system's effectiveness and efficiency can only be possible through repeated experiences. The following steps show a logical sequence of development; however, any step can be developed with the rest if the situation is more convenient for such parallel development:

1. Decide the major objectives or terminal outcomes of the course.
2. Roughly break down these objectives or outcomes into as many components as manageable to monitor student progress.
3. Decide how often to assess progress throughout the course duration for each component. This assessment may mainly be quantitative, but qualitative assessment (e.g., observational log and portfolio) can be a good supplement.
4. Decide the data collection method to measure the effectiveness of remedial activities following each assessment. Keep descriptive records of activities and their outcomes (e.g., log keeping and audio or video recording) in addition to quantitative data automatically obtained through regular assessment.
5. Apply the results of #1 through #4 to language course development. Repeat and refine the system of assessment and data-collection.
6. Prepare the technology infrastructure (e.g., install software and organize the database).
7. Maximally transform paper-and-pencil versions of assessment into technology-based formats.
8. Develop a technology system that analyses data and displays results.
9. Make the whole learning/teaching system meaningful and user-friendly.

We should not focus on large-scale goals or attend excessively to small details in the initial stages of development. It is wiser to start with the simplest case and repeat the implementation cycle many times with continuous increments of elaboration. Action research provides the most useful framework to organize such a cyclic development process. The process must also be flexible so as not to endanger language educators' most important responsibility – teaching.

The final section introduces an example in which the authors are implementing TAADLS in a Japanese language course.

4.2 An example of implementation of TAADLS: Cycle one

The idea to employ adaptive and dynamic functions in language programs occurred to the authors in the fall of 2000 when discussing collaborative research with language course students as well as colleagues. The students showed strong interest in this project, which led to two group meetings in December 2000 that solidified the mutual commitment to pursue this idea. During spring 2001, occasional meetings guided the two major projects over summer 2001. One psychology major senior took responsibility for developing a mini-curriculum to teach two sets of *kana* letters and 30 *kanji* (Chinese characters) to elementary Japanese language students. Her summer development was paid by the EXCEL scholar program of Lafayette College. The author also arranged a grant-based team to build the infrastructure of a technology system. Two computer science major students worked full-time for this part of the project during summer 2001, paid through a mini-grant of the Mellon Foundation, provided by the Centre of Educational Technology at Middlebury College, Vermont.

The fall semester course of elementary Japanese has employed the new mini-curriculum developed over the summer. Two additional advanced language students will join this project during fall to support its research and implementation. For the technology part, one faculty member of the computer science department joined the technological development of the project in August 2001. Under his guidance, three computer major students will transform paper-and-pencil materials into a web-based format and organize a system that records student performance and analyses data.

This first cycle is the beginning of a long term research and development project, which requires various kinds of collaboration among multiple institutes as well as within Lafayette College. This project has begun in a little developed area of the current Japanese language curriculum. Later, it will expand to other language skills as well as to more advanced courses. The authors hope that more practitioners will start collaborative development of a highly integrated language learning system based on technology.

References

Argyris, C., Putnam, R., & Smith, D. M. (1985). *Action science.* San Francisco: Jessey-Bass.

Ariizumi, Y. (1998). *Five ways of knowing action research.* Unpublished dissertation, Brigham Young University, Provo, UT.

Bergstrom B. A., & Lunz, M. E. (1999). CAT for certification and license. In F. Drasgow, & J. Olson-Buchanan (Eds.), *Innovations in computerized assessment.* Mahwah, NJ: Lawrence Erlbaum.

Bolig, E. E., & Day, J. D. (1993). Dynamic assessment and giftedness: The promise of assessing training responsiveness. *Roeper Review, 16*(2), 110-3.

Budoff, M. (1987). Measures for assessing learning potential. In C. S. Lidz (Ed.), *Dynamic assessment: An interactional approach to evaluating learning potential* (pp. 173-95). New York: Guilford Press.

Burns, M. S. (1985). *Comparison of "graduated prompt" and "mediational" dynamic assessment and static assessment with young children (Tech. Rep. No. 2)*. Nashville, TN: Vanderbilt University, John F. Kennedy Centre for Research on Human Development.

Campione, J. C. (1989). Assisted assessment: A taxonomy of approaches and an outline of strengths and weaknesses. *Journal of Learning Disabilities, 22,* 151-65.

Campione, J. C., Brown, A. L., Ferrara, R. A., Jones, R., & Steinberg, E. (1985). Breakdowns in flexible use of information: Intelligence-related differences in transfer following equivalent learning performance. *Intelligence, 9,* 297-315.

Carslon, R. (1994). Computer-adaptive testing. A shift in the evaluation paradigm. *Journal of Educational Technology Systems, 22,* 213-24.

Chalhoub-Deville, M., & Deville, C. (1999). Computer-adaptive testing in second language contexts. *Annual Review of Applied Linguistics, 19,* 273-99.

Cioffi, G., & Carney, J. J. (1983). Dynamic assessment of reading disabilities. *Reading Teacher, 36*(8), 764-8.

Cochran-Smith, M., & Lythe, S. (1993). *Inside/outside: Teacher research and knowledge.* New York: Teachers College Press.

Day, J. D., & Hall, L. K. (1987). Cognitive assessment, intelligence, and instruction. In J. D. Day & J. G. Borkowski (Eds.), *Intelligence and exceptionality: New directions for theory, assessment, and instructional practices* (pp. 57-80). Norwood, NJ: Ablex.

De Ayala, R. J. (1992). The nominal response model in computerized adaptive adaptive testing. *Applied Psychological Measurement, 16*(4), 327-43.

Delandshere, G. (1996). From static and prescribed to dynamic and principled assessment. *The Elementary School Journal, 97*(2), 105-20.

Delclos, V. R., Burns, M. S., & Kulewicz, S. J. (1987). Effects of dynamic assessment on teachers' expectations of handicapped children. *American Educational Research Journal, 24*(3), 325-36.

Dixon-Krauss, L. (1996). *Vygotsky in the classroom: Mediated literacy instruction and assessment.* London: Longman.

Drasgow, F., & Olson-Buchanan, J. B. (1999). Beyond bells and whistles: An introduction to computerized assessment. In F. Drasgow, & J. B. Olson-Buchanan (Eds.). *Innovations in computerized assessment.* Mahwah, NJ: Lawrence Erlbaum.

Drasgow, F., & Olson-Buchanan, J. B. (Eds.), (1999). *Innovations in computerized assessment.* Mahwah, NJ: Lawrence Erlbaum.

Dunkel, P. (1999). *Considerations in developing and using computer-adaptive tests to assess second language proficiency* (ERIC Digest). (ERIC Document Reproduction Service No. ED 435 202)

Elliott, J. (1987). Educational theory, practical philosophy and action research. *British Journal of Educational Studies, 35*(2), 149-69.

Feuerstein, R. (1979). *The dynamic assessment of retarded performers: The learning potential assessment device, theory, instruments, and techniques.* Baltimore, MD: University Park Press.

Feuerstein, R. (1980). *Instrumental enrichment: An intervention program for cognitive modifiability.* Baltimore, MD: University Park Press.

Feuerstein, R., Miller, R., Hoffman, M. B., Rand, Y., Mintzker, Y., & Jensen, M. R.

(1987). Cognitive modifiability in adolescence: Cognitive structure and the effects of intervention. In M. Heiman, & J. Slomianko (Eds.), *Thinking skills instruction: Concepts and techniques.* Washington, DC: National Education Association.

Feuerstein, R., Rand, Y., Hoffman, M. B., Egozi, M., & Shachar-Segev, N. M. (1991). Intervention programs for retarded performers: Goals, means, and expected outcomes. In L. Idol & B. F. Jones (Eds.), *Educational values and cognitive instruction: Implications for reform* (pp. 139-78). Hillsdale, NJ: Erlbaum.

Fogarty, R. (1998). *Balanced assessment: K-college.* (ERIC Document Reproduction Service No. ED 426 059)

Gottman, J. M. (1981). *Time-series analysis: A comprehensive introduction for social scientists.* Cambridge: CUP.

Haywood, H. C. (1992). Interactive assessment as a research tool. *Journal of Special Education, 26*(3), 253-68.

Henning, G., Hudson, T., & Turner, J. (1985). Item response theory and the assumption of unidimensionality for language tests. *Language Testing, 2,* 141-154.

Jitendra, A. (1993). Dynamic assessment as a compensatory assessment. *Remedial and Special Education, 14*(5), 6-18.

Kim, H., & Plake, B. S. (1993, April). *Monte Carlo simulation comparison of two-stage testing and computerized adaptive testing.* Paper presented at the Annual Meeting of the National Council on Measurement in Education, Atlanta, GA.

Kingsbury, G., & Houser, R. (1993). Assessing the utility of item response models: Computer adaptive testing. *Educational Measurement: Issues and Practice, 12,* 21-7.

Kletzien, S. B., & Bednar, M. R. (1990). Dynamic assessment for at-risk readers. *Journal of Reading, 33*(7), 528-33.

Lewin, K. (1946). Action research and minority problems. *Journal of Social Issues, 2*(4), 34-46.

Lidz, C. S. (1992). Dynamic assessment: Some thoughts on the model, the medium, and the message. *Learning and Individual Differences, 4*(2), 125-36.

Linacre, J. M. (1999). Understanding Rasch measurement: Estimation methods for Rasch measures. *Journal of Outcome Measurement, 3*(4), 382-405.

Lord, F. M. (1980). *Applications of item response theory to practical testing problems.* Hillsdale, NJ: Lawrence Erlbaum.

Luther, M. (1982). The Feuerstein method with ESL students. *TESL Talk, 13*(2), 69-73.

Markham, K. (1993). *Standards for student performance* (ERIC Digest, No. 81). (ERIC Document Reproductive Service No. ED 346 082)

Minick, R. D., & Medlin, S. M. (1983). Anticipatory evaluations in HRD programming. *Training and Development Journal, 37*(5), 89-94.

Noffke, S. E. (1994). Action research: Towards the next generation. *Action Research, 2*(1), 9-21.

Pascal, C. (1995). *Effective early learning: Mapping diversity and tracking development.* Paper presented at the European Conference on the Quality of Early Childhood Education, September, Paris. (ERIC Document Reproductive Service No. ED 390 534)

Pendlebury, B. (1985). Feuerstein and the FE curriculum. *British Journal of Special Education, 12*(1), 13-5.

Presseisen, B. Z., Sternberg, R. J., Fischer, K. W., Feuerstein, R., & Knight, C. C. (1990). *Learning and thinking style: Classroom interaction.* Washington, DC: National Education Association.

Roos, L. L., Wise, S. L., & Finney, S. J. (1998, April). *Comparing restricted and unrestricted self-adaptive testing as alternatives to computerized adaptive testing.* Paper presented at the Annual Meeting of the National Council on Measurement in Education. San Diego, CA.

Rosenfield, S., & Nelson, D. (1995). *The school psychologist's role in school assessment* (ERIC Digest). (ERIC Document Reproductive Service No. ED 391 985)

Stocking, M. L. (1994). *An alternative method for scoring adaptive tests* (Rep. No. RR-94-48). (ERIC Document Reproduction Service No. ED 380 498)

Straetmans, G. J. J. M., & Eggen, T. J. H. M. (1998). Computerized adaptive testing: What it is and how it works. *Educational Technology, 38*(1), 45-52.

Swanson, H. L. (1996). Classification and dynamic assessment. *Focus on Exceptional Children, 28*(9), 1-20.

Taylor, R. (1993). Assessment in the nineties. *Diagnostique: Professional Bulletin of the Council for Educational Diagnostic Services, 18,* 113-122.

van Manen, M. (1990). Beyond assumptions: Shifting the limits of action research. *Theory Into Practice, 29*(3), 152-7.

Vye, N., Burns, M. S., Delclos, V. R., & Bransford, J. D. (1985). *Dynamic assessment of intellectually handicapped children* (Tech. Rep. No. 4). Nashville, TN: John F. Kennedy Centre for Research on Education and Human Development.

Wainer, H. (1990). *Computer adaptive testing: A primer.* Hillsdale, NJ: Erlbaum.

Way, W. D. (1997). *Protecting the integrity of computerized testing item pools.* Paper presented at the annual meeting of the National Council on Measurement in Education, Chicago, IL.

Whitehead, J. (1989). Creating a living educational theory from questions of the kind, "How do I improve my practice?" *Cambridge Journal of Education, 19*(1), 41-52.

Wise, S. L. (1999, April). *Comparison of stratum scored and maximum-likelihood scored CATs.* Paper presented at the Annual Meeting of the National Council on Measurement in Education. Montreal, Quebec, Canada.

Zuber-Skerritt, O. (1992). *Action research in higher education – Examples and reflections.* London: Kogan Page.

5

MI Theory and CALL: Personalized Education and Learning for Understanding

Laurence M. Dryden
Nagoya Institute of Technology
Michelle Henault Morrone
Nagoya University of Foreign Studies

1. Introduction

Multiple intelligence (MI) theory and CALL developed separately but simultaneously from the late 1970s onward, arising respectively from major breakthroughs in cognitive psychology and the increasing use of personal computers. Taken together with fundamental changes in the nature of literacy and learning during the same period, it is not surprising that the disciplines of MI theory and CALL have converged to serve the interests of educational reform. Nevertheless, both MI theory and CALL are pedagogically neutral and need to be activated by specific pedagogical goals. Among those goals, as recommended by experts cited throughout this article, are the individualisation of learning and the promotion of education for deep understanding.

Educators who use computers generally agree that personal computers provide powerful tools for students to learn languages and appreciate cultures. What is less well understood, however, is that much of the power of CALL to stimulate learning can be explained by MI theory (Dryden, 1998). Indeed, grasping the relation of MI theory and CALL is particularly helpful in comprehending the principal educational issues of our time. Contemporary education stands at a turning point, poised between a choice of guiding metaphors: The long-established twentieth-century models of mass production and standardisation, as seen in uniform instruction, contrasted with the shifting landscape of the twenty-first century world in which post-modern, post-industrial digital communication is giving rise to new forms of literacy that require new approaches to learning. Drawing another metaphor from the realm of physics, Negroponte (1998) frames the contrast of the industrial age and the post-industrial age in terms of "atoms" versus "electrons" (p. 21). In effect, the computer-based communication revolution has led us into completely new territory in all areas of modern culture, including education. The convergence of MI theory and CALL offers one way to help students through the difficult, unfamiliar terrain that must be negotiated in this time of rapid transition.

2. MI theory and CALL in the search for alternatives in education

MI theory is now generally accepted as part of the atmosphere of American K-12 education and has attracted the interest of many academics in higher education as well as educators at all levels around the world. Even its critics have acknowledged its growing and pervasive influence (Gardner, 1999, p. 100). From the late 1970s, MI theory was pioneered by Howard Gardner, a cognitive psychologist in the Graduate School of Education at Harvard University. Gardner (1983, 1985) explains that one early incentive to investigate the nature of human intelligence arose from observing that standardized IQ tests did a reasonably good job of predicting success in school but were much less reliable in foretelling success in life beyond formal schooling (Dryden & Morrone, 1999, p. 27). In effect, Gardner noticed that those who do well in school do not always do so in life, and, conversely, some students who do badly in school, often branded as "problem children" or "underachievers," may thrive once free of the constraints of formal education.

In Japan, the situation is more complex, and suffering through the throes of the system proves one worthy to be among the elite and, as such, to find "Japanese happiness" (choices in career and marriage, and freedom to select from a cornucopia of varied tastes), and comfort (a relaxed but honourable position, job security, social acceptance and connections, and a generally full range of life chances). While not hard to convince most people that individual success outside formal education is possible, it is more difficult to explain that one can still experience complete satisfaction even without societal appraisal. What Gardner calls *interpersonal* intelligence, while not formally part of school, might be said to guide the hidden curriculum in Japan, in that without it one can achieve very little. Brains or skill alone may provide a good life but will not propel one to the top of one's field.

From years of observing American schools and hospitals, Gardner (1983, 1985) concluded that general intelligence (IQ) tests, accepted for most of the twentieth century as reliable measures of human potential, are seriously flawed. By reducing the measurement of intelligence to pencil-and-paper exams with discrete-point questions and single-right answers, such tests encompass too narrow a range of cognitive abilities and rest on too limited a model of the human mind. Gardner also concluded that intelligence is not unitary but pluralistic: That is, throughout history, humans have shown many ways of being intelligent in particular cultural contexts. Twentieth-century schools in industrialised countries, however, emphasised the linguistic and logical-mathematical intelligences because of their value and utility in modern bureaucratic culture. The were often readily measurable on standardised test that developed as gate-keeping instruments during the same period.

The implications of MI theory for educational policy and practice are profound and far-reaching. From Gardner's point of view, the growing preference in America for standardized testing (a long-established practice in Japan) favours an inauthentic and unreliable means of measuring the full range of each individual's true potential (Dryden, 1998, p. 96). A common-sense idea of "success" exists in all cultures, and pen-and-paper appraisal of intelligence is a modern and rather artificial development. With the help of fresh thinking like Gardner's, contemporary education may eventually overcome prejudices in this area. In America now, there is much concern about the excessive emphasis on testing. For example, *The American School Board Journal* (August, 2001) reports on students and parents protesting the Bush administration's idea to "test every child, every year."

Computers, while widely used for standardized testing, are also being put to more imaginative uses. CALL, as a powerful array of tools for learning, holds great potential for working together with MI theory to support a host of progressive goals in language learning – from empowering marginalised learners to making learning more satisfying for all. The convergence of CALL and MI theory also promises to cultivate real skills and interests that students can demonstrate their ability in forums more authentic than standardised tests.

3. Intelligence defined, with special relevance to CALL

MI theory developed from Gardner's dissatisfaction with the ways that schools have mismeasured intelligence and have often stunted human potential. Gardner (1999) says that in his earliest thinking about MI, he found it useful to define intelligence as "the ability to solve problems or to create products that are valued within one or more cultural settings" (p. 33). Later, he refined the definition, conceptualizing an intelligence as considerably more than an ability, amounting to "a biopsychological potential to process information that can be activated in a cultural setting to solve problems or create products that are of value in a culture" (pp. 33-34). With this definition, Gardner meant to account for cognitive activity – *neural potentials* that may be activated through interaction with culture and individual experience – across the wide range of domains in which humans work and play, perform tasks, and make achievements, inventions, and discoveries. Gardner's definition of intelligence has far-reaching implications for a better understanding of CALL: The personal computer is an immensely versatile tool for solving problems and creating products of value in many walks of contemporary life, making its role in education increasingly valued and necessary.

4. The eight (or more) intelligences and CALL

Gardner (1983) initially identified seven distinct intelligences: the *school* intelligences, linguistic, and logical-mathematical; the *artistic* intelligences, musical, visual-spatial; the *personal* intelligences, interpersonal and intrapersonal; and one hybrid intelligence, both *artistic* and *physical*, termed *bodily-kinesthetic*. Gardner conceded that other intelligences probably existed and might eventually be confirmed through empirical testing guided by criteria he had established–among them, evidence from brain research, human development, evolution, and cross-cultural comparisons (Gardner, 1993, pp. 26-27). (For technical elaboration of these criteria for an intelligence, see Gardner, 1993, pp. 15-16; Armstrong, 1994, pp. 3-11; and Dryden, 1998, p. 97.) Following further research, Gardner (1999) acknowledged an eighth intelligence, which he called naturalist (pp. 48-52).

As the present study aims to show, CALL engages all eight intelligences through the many ways that learners can work with computers. Essentially, multimedia and online activities permit learners, including language learners, to make choices about what they will learn, at what pace, and in whatever media and modes of intelligence are best suited to their individual needs. Support through many different "media of intelligence" – graphics, video, animation, and sound, as well as text – can help language learners make gains in learning, while offering valuable concessions to the majority of learners for whom linguistic intelligence is not their strongest suit.

[Note: The following discussion of MI theory synthesizes ideas from Armstrong (1994, pp. 2-3); Gardner (1983, 1985, pp. 73-276); Gardner (1993, pp. 17-26); and Gardner (1999, pp. 48-54). The interspersed discussions of MI and CALL derive from Thornburg (1993), Armstrong (1994, pp. 158-162), Dryden (1998, 2000), Dryden and Morrone (1999), and Gardner (2000).]

4.1. The school intelligences

These are the intelligences (verbal and mathematical) measured on most standard IQ tests commonly administered in U.S. K-12 schools and on college placement tests (e.g., the SAT and the GRE). In Japan, these two intelligences also dominate the educational system and are enshrined in the elaborate series of entrance exams that serve as high-stakes gate-keeping rites of passage for Japanese students from junior high (and sometimes even earlier) through university. They are the intelligences most highly valued in twentieth-century schools in industrialised countries, and they reflect the cultural values and perceived educational needs of that age. (Japan, as usual, is somewhat exceptional in the industrialized world: Its education often falls short in MI logical intelligence, even though Japanese students score highly in mathematical pencil-and-paper tests. The Western idea of "logic" may be culturally biased.)

The school intelligences are extremely important, but MI theory holds they are not the only ones worth acknowledging and cultivating. Gardner (1999) wryly notes that the "language-logic mind" is what one would expect to find epitomized by a law professor (p. 150). Imposition of such a frame of mind on all students is hardly an appropriate goal for a democratic system of education that should more properly meet students where they are and accommodate their diverse needs.

Linguistic intelligence involves the ability to use words skilfully in speaking and in writing, as well as sensitivity to the sounds, structure, meaning, and functions of words and languages. Linguistic skills include giving and remembering information; persuading others to accept another point of view; explaining, teaching, and learning; talking about language itself, and writing novels and poetry. In CALL, linguistic intelligence is processed through text display and text input, the principal – but by no means only – way of communicating with others and using computers to solve problems and create products through language.

Logical-mathematical intelligence involves the abilities to use numbers skilfully and reason well, solve puzzles easily, and follow long chains of reasoning. Logical-mathematical skills include understanding the qualities of numbers and principle of cause and effect, as well as the ability to predict what will happen or explain what has happened. In CALL, students employ logical-mathematical intelligence when using computational tools and search methods. While computer programmers operate at the high end of logical-mathematical intelligence, language learners can use this intelligence to find the solutions to a variety of questions about language and culture by means of web searches and information retrieval.

4.2 The artistic intelligences

MI theory is especially amenable to those who work in or appreciate the fine and performing arts, as it raises artistic accomplishment and creativity to a status comparable to that of the school intelligences. Gardner has argued consistently that a proper model of the mind needs to account for the full range of human cognition. Moreover, education should concern itself with more than preparation for employment; it should also equip students for an appreciation and enjoyment of life, particularly of the arts.

Musical intelligence involves the abilities to produce and appreciate rhythm, pitch, and melody; appreciate the forms of musical expression, recognizing songs and longer musical works; vary speed, tempo, and rhythm in melodies, and compose music. In CALL, students activate this intelligence in musical recordings and playback of sounds. They can download songs (or create them) and use them to develop a sense of the rhythm of language, learn and remember vocabulary, or simply gain a better aesthetic and personal appreciation of language.

Visual-spatial intelligence involves sensitivity to form, space, colour, line, shape, and depth; the abilities to perceive the visual world accurately and to represent visual or spatial ideas, either graphically or mentally in three dimensions. In CALL, students engage with visual-spatial intelligence through graphics and digitised images. A visual complement to verbal-linguistic text can be a powerful support for language learning, as the visual image conveys meaning that augments and sometimes surpasses the written word itself.

4.3 The personal intelligences

Many might be surprised – and others would dispute – that personal skills should be elevated to the level of intelligence, yet these abilities are highly valued and rewarded in daily life and in the modern corporate and political worlds. They meet the test of Gardner's criteria and account for otherwise anomalous cultural phenomena. For example, two recent American presidents lacking academically, i.e., in the school intelligences, were nonetheless raised to the summit of U.S. political power by virtue of their interpersonal intelligence strengths.

Interpersonal intelligence is the means by which individuals understand and skilfully respond to others' moods, feelings, motivations, desires, and intentions. It helps a person work effectively, get along well with others, and persuade them to do what he or she asks of them. In CALL, students use interpersonal intelligence through group interaction involving learner-created (or collaborative) multimedia presentations, email, and online discussion lists.

Intrapersonal intelligence, by contrast, is the means by which people understand themselves, and come to know their strengths and weaknesses; discriminate among moods, desires, feelings, emotions, and intentions; understand how they are similar to or different from others; remind themselves to do something, and know how to control their feelings. In CALL, students use intrapersonal intelligence for self-directed exploration, navigating content any way they wish.

4.4 The physical intelligences

Often relegated to the lowly status of "manual labour," the *physical* intelligences also satisfy Gardner's criteria. Gardner (1999) traces the snobbishness about physical activity to the Cartesian split between mind and body (p. 96). In fact, prejudice against the physical intelligences can be traced to classical antiquity, when manual labour was relegated to slaves and artisans and intellectual work considered the proper vocation of free men and aristocrats. By validating the physical intelligences, MI theory does much to restore a balanced view of these areas of human cognition.

Bodily-kinesthetic intelligence can be seen in the ability of people to use their bodies to express ideas and feeling through movement, as well as the capacity to handle objects skilfully. This intelligence involves physical coordination, flexibility, speed, and balance. In CALL, students activate bodily-kinesthetic intelligence whenever they work with online animated images and videos.

Naturalist intelligence involves the ability to recognize and classify the numerous species of flora and fauna in the environment, as well as the capacity to care for, tame, or interact subtly with living creatures. In CALL, students use naturalist intelligence when they hunt for, gather, and classify information.

5. MI theory and CALL in joint practice

The convergence of MI theory and CALL offers hope for many needed reforms in education. Numerous educational computer applications in general, and many CALL applications in particular, clearly engage multiple intelligences as part of their appeal for learners. CALL provides an array of options for learners and teachers to use in customizing the learning process, tailoring activities to particular individual needs and interests, as MI theory requires. Furthermore, MI theory and CALL, while essentially both pedagogically and morally neutral, can work together to serve a variety of specific educational goals, stretching across a vast range of desired outcomes for beginning, intermediate, and advanced language learners.

Computer-based learning departs from conventional text- and lecture-based instruction by engaging all the multiple intelligences, and giving learners choices of content, media, and pace of learning. It does this through hypertextual linking of text, sound, images, and motion on videodiscs and the Internet, integrating and conveying information to learners in mutually-supportive media. In this way, CALL software serves a wide range of learners by enabling them to make use of a larger repertoire of cognitive skills than conventional teaching and learning permit, thereby accommodating even those students whose cognitive strengths have traditionally been neglected.

5.1 MI theory and computers: Reforming education by making it personal

MI theory helps in analysing many problems of modern education and also suggests possible solutions. A major problem is that the "big three" intelligences – linguistic, logical-mathematical, and to a lesser extent intrapersonal – dominate most curricula; by contrast, the other five intelligences receive nominal attention, usually in "extra-curricular" activities or in art and music programs that are invariably the first budget items cut in hard times. (The generalization about the big three intelligences holds true for schools in North America and Europe. By contrast, in Japanese junior and senior high schools, with their social emphasis on group identity, interpersonal intelligence displaces intrapersonal intelligence, though in other respects the school intelligences of language and mathematics remain paramount.) This imbalance was tolerated throughout the twentieth century but may not endure much longer: The world outside school continues to change rapidly, driven by the ICT revolution in personal computing and, more recently, in portable communication devices virtually universal among young Japanese today.

The discrepancy between the world of students and the world of school shows itself in many significant ways. Most teachers tend to score highly in one or more of the big three intelligences, while most students incline toward the other five – a telling sign of the mismatch between typical school curricula on one side and the needs, interests, and abilities of most learners on the other. Moreover, the places in which students immerse themselves when not in school – contemporary mass media, including and perhaps pre-eminently computer environments – have become increasingly visual, while traditional school curricula remain tied to the printed word. Such widening discrepancies between the world of students and that of school may have contributed to soaring increases in school dropout rates and classroom discipline problems, particularly in Japan, a fact which may surprise many Western readers.

MI theory also serves in the analysis of another major problem in modern schooling, the uniform curriculum (which, while generally devised in the interests of equality, often produces inequality). Modern education rests on an industrial model of standardization of parts and mass production. The resulting Procrustean bed of contemporary schooling has truncated the minds and spirits of many students, or else stretched them unhealthily, in the interest of making everyone fit a uniform model. By contrast, Gardner (1999) regards MI theory as a "ringing endorsement of three key propositions: We are not all the same; we do not all have the same kinds of minds (that is, we are not all distinct points on a single bell curve); and education works most effectively if these differences are taken into account rather than denied or ignored. Taking human differences seriously lies at the heart of the MI perspective" (p. 91).

MI theory recommends that learning be customized, based upon the understanding that each learner is cognitively unique. Gardner's (1999) affirmation that everyone has a unique blend of intelligences leads to "the most important implication of the theory for the next millennium. We can choose to ignore this uniqueness, strive to minimize it, or revel in it . . . . I suggest that the biggest challenge facing the deployment of human resources is how best to take advantage of the uniqueness conferred on us as the species exhibiting several intelligences" (p. 45). Computers offer vast resources for making the most of human uniqueness. The question is whether educators can find the imagination and the will to use them for this purpose.

In Japan, for example, proponents of educational reform are calling specifically for more *creativity* and *individuation* (individualisation) in the schools (Ministry of Education, 1997, 2001). These buzzwords are echoed in other Asian countries, as governments throughout the region have pinned their hopes on new educational practices which, in time, may help revive their stagnant economies (Dryden, 2000; Mc-Murray, 1998). In Japan, there is also a growing recognition of the global links between education and work. Educational criteria now require skills to be real: It is no longer enough to get into college and then sit back and chat with friends during a four-year vacation. In the same vein, the graduate school boom in Japan means that education is becoming more competitive – real skills and abilities are now valued, suggesting that Gardner's MI theory should play a bigger role in the official curriculum. Although not openly recognized, the researchers behind the ideas that push the Japanese Ministry of Education in the direction of reform are well aware of developments in this area in other countries, especially the US (Ministry of Education, 1997, 2001).

In Japan, however, individuality and creativity have traditionally met with skepticism to say the least. Such attitudes are generally ascribed to Japan's Confucian heritage, which is thought, by virtue of its reverence for hierarchy and tradition, to stand in the way of creativity. There would seem to be little room for intellectual flexibility in solving problems and creating products (i.e., Gardner's definition of an intelligence). Nevertheless, as Dryden (2000) observes, the Confucian roots of Japan (as well as of China and Korea) are actually grounded in *self-cultivation* and *learning for oneself* (de Bary, 1997). As they seek to reinvent themselves, Japan and other East Asian countries may return to the origins of Confucianism and bring about a greater emphasis on *self* and a balance between individual and social responsibility. If so, as Dryden (2000) notes, "The native Confucian tradition holds the potential for transforming the current educational systems in East-Asian countries and promoting the kind of creative and innovative thinking that the Asian economies require for their own regeneration" (p. 113). Among the tools available for recasting education as self-cultivation are MI theory and CALL – ideally working together in harmony.

Reshaping public education in Japan will not be easy, however, and the endeavour is handicapped by a lag of over ten years behind the West – the place from which Japan borrowed the mass-production model of education that no longer works very well anywhere. Customized learning as recommended by MI theory, while appealing, remains problematic. How, one might ask, can the schools, with their limited resources, manage to do better for individuals than they have already tried and not consistently succeeded to do for great masses of students?

Computers, of course, represent great promise in customizing education, and are currently being introduced into Japanese public education with the speed and blind faith seen in American schools over a decade earlier. As the American experience showed, however, if computers are to have any significant impact on education, the human side of computer-based learning will have to be emphasized. Within schools, as technology continues to be introduced into the curriculum, the human factor is crucial in helping people become computer literate: In this regard, the principle of "high tech, high touch" is essential. When trying to create links between school and the wider community, one must accept that society has already changed so much that we cannot return to the good old days of neighbourhoods linked by conversations on summer porches. The contemporary equivalent may be found in new virtual communities created by email, discussion lists, and other computer-based resources that involve human interaction through personal communication.

In effect, interpersonal intelligence must operate for computer-based education to become truly engaging. The danger, however, is that even with computers, learning and information gathering may simply follow traditional teaching forms which emphasize verbal and logical intelligence over others. In Japan, where computers are being introduced into junior and senior high schools on a wide scale in 2002, the first impulse of many administrators and teachers is likely to be to use computers to standardize learning on a mass scale – a major failure to exploit the full potential of computers in education. (Computers have been used for language teaching and other purposes in Japanese universities for over a decade, but even there the record is mixed.)

How can the multiple intelligences be effectively addressed through technology? A recent study by Harold Hodgkinson at Harvard (Arenson, 2001) revealed that American college students, discussing happy and unhappy experiences of college, insisted that "personal engagement" was a crucial part of a successful education. Similarly, if computers are to enrich our lives, they must serve the needs of both inter- and intra-personal intelligences; otherwise, we are doomed to become, as Thoreau (1854) warned, "the tools of [our] tools."

Adding further to the complications of educational reform, however, the post-industrial world now places greater demands on schools and graduates than ever before. Dede (2000) argues that aiming for higher standards in the current curriculum will not be sufficient to prepare students for life in the twenty-first century:

> Students – and teachers – need to master new skills that the current curriculum may not address, skills that were not central to the industrial society of the past century but are vital to the knowledge-based economy of the new one. These skills include: the ability to collaborate with diverse teams of people – face to face or at a distance – to accomplish a task; to create, share, and master knowledge by assessing and filtering quasi-accurate information; and to "thrive on chaos," that is, to make rapid decisions based on that incomplete information to resolve to [sic] novel dilemmas. (p. 171)

In short, urgent demands for new kinds of literacy in thought and action are rapidly evolving in the post-modern world, and simply doing more of the old-school curriculum will not equip students for the world they are going to live and work in. As Dede (2001) observes, educators can no longer afford to continue with the classroom charade, "We pretend to teach. They pretend to learn." Nothing short of a fundamental reinvention of education will do.

Gardner and Dede agree that one way out of the dilemmas faced by contemporary education is found in new technologies, specifically those made possible by personal computers. Used appropriately, computer-based learning can serve Gardner's goals of providing individualized learning and expanding beyond the intelligences traditionally honoured by the schools, as well as mediating the new kinds of literacy that

Dede envisions. Nord (1997) hails the recent developments in CALL software as a "Copernican revolution in education, a paradigm shift from the nineteenth-century model of teacher and classroom centred education, to the twenty-first-century model, with the centre occupied by learners themselves." Moreover, Dryden (1998) explains that CALL tutorial software supports this new model by maximizing student control over the learning process, while CALL tools enable learners to create and communicate with ease, allowing use of the target language for real purposes in relatively risk-free environments: "Beyond providing more choices for learners, the best CALL software also appeals to a wide range of human intelligences, making language learning more successful through visual and auditory support for text comprehension" (p. 100). The next section will consider this vital and constructive relationship of MI theory and computer-based learning, with particular emphasis on CALL.

5.2 MI theory and CALL: Activities across a continuum

Beyond Gardner's insistence that any MI-based curriculum should serve individual uniqueness, MI and CALL have no ready-made, prescribed uses and can therefore be enlisted for any number of pedagogical purposes. Gardner (1999) describes MI theory as "a handmaiden of good education, once educational goals have been established on independent grounds" (p. 166). Much the same could be said of computer-based learning. Both MI and CALL are best thought of as tools for learning, to be taken up and used once specific curricular goals have been defined.

The question of educational goals, then, becomes central to the discussion of what MI theory and CALL can be invoked to serve. What is it, exactly, that teachers want their language learners to accomplish? Traditionally, language proficiency levels have been divided into beginner, intermediate, and advanced, and conventional division of the curriculum includes reading, writing, speaking, and listening. Beginners can profit from extra-linguistic support through talking books and other resources on CD-ROM, DVD, and the Internet that integrate text, pictures, and the spoken word, as well as songs and video. Both beginners and intermediate students can make use of extensive databases of online learning activities on websites such as the *Internet TESL Journal* (Kelly & Kelly, 2002). Based in Aichi Prefecture, Japan, this internationally-renowned online resource for EFL learners allows students to select from thousands of quizzes, readings, and listening activities, work at their own pace and level of interest, and receive instant feedback. Learners can also follow links to sites specially designed to help them find other EFL students with whom to correspond and collaborate, including *Student Lists* (Robb, 2002) and *Dave's ESL Café on the Web* (Sperling, 2002).

5.3 MI theory and CALL: Learning for understanding

For high-intermediate and advanced learners, however, the goals might rise to levels approaching those appropriate for native speakers. Gardner (1999) argues that the proper goal for learning should be nothing less than understanding in depth: "Education in our time should provide the basis for enhanced understanding of our several worlds–the physical world, the biological world, the world of human beings, the world of human artifacts, and the world of the self" (p. 158). Gardner, then, affirms the value of content-based and theme-based learning, which language learning, as well as other disciplines, should serve. He continues:

Note that this goal does not mention the acquisition of literacy, the learning of basic facts, the cultivation of basic skills, or the mastery of the ways of thinking of the disciplines – achievements that should be seen as means, not ends in themselves. We learn to read, write, and compute not so that we can report these milestones (as we might post an attendance record), nor even so that we can achieve a certain score on an admissions test. Rather, literacies, skills, and disciplines ought to be pursed as tools that allow us to enhance our understanding of important questions, topics, and themes . . . . Resources invested in formal education can best be justified if, at the end of the day, all students show enhanced understanding of the important questions and topics of their time. And once such a commitment has been made, the powerful ideas of multiple intelligences can be mobilized to achieve that goal. (1999, p. 159)

The efforts to achieve basic literacy and fluency in language study are thus properly thought of as stepping-stones to higher cognitive objectives, i.e., the ability to understand the world in depth and function in it more effectively. What topics, then, would be appropriate to ask learners of a foreign language to understand in depth? Any number of themes might be possible. Understanding of themselves is a good starting point. Having students deal with the issue of identity through the question of "Who am I?" is intrinsically motivating and meets the requirement of an MI-based curriculum that learning be personalized (Dryden & Morrone, 1999, pp. 30-31). Understanding of other cultures – their social systems and values, artistic achievements, and history – is a desirable goal, too. For language learners in special programs or departments such as science and technology, business, or any of the liberal arts, topics in their own domains would be worthwhile to learn about and gain a deep understanding of. As Gardner (2000) insists, "We should help students to understand the major ways of thinking that have developed in the disciplines. Here is where the new technologies can really come into their own" (p. 34).

To help achieve this, MI theory and CALL together can enrich students' understanding of themes and topics in their disciplines by providing tools for cognition, communication, research, and reflection. Gardner (2000) observes, "With new software and the World Wide Web, it is possible to receive and to manipulate all kinds of (hopefully accurate) data, captured in a wide range of symbol systems, and evaluate respective claims and counterclaims." By making use of the extensive resources on CD-ROMs, videodiscs, and the Internet, students can, according to Gardner "easily access the relevant documents, including photographs or film records, create one's own models, evaluate them in the light of contemporary and retrospective accounts, and relate them to events in today's newspaper." Gardner acknowledges that such studies could certainly be carried out without benefit of multimedia and cybercommunications, but he sees in their intrinsically motivating power the potential to energize the learning process: "The new technologies make the materials vivid, easy to access, and fun to play with – and they readily address the multiple ways of knowing that humans possess" (pp. 34-35).

The convergence of MI theory and CALL serves not only in searching for and accumulating information and knowledge for understanding, but also in presenting that understanding to others. Gardner (1999) insists that understanding is facilitated best when one is required to present one's insights in a public forum:

> When it comes to understanding, the emphasis falls properly on performances
> that can be observed, critiqued, and improved . . . . Instead of "mastering con-
> tent," one thinks about the reasons *why* a particular content is being taught and
> how best to display one's comprehension of that content in a publicly accessi-
> ble way. When students realize they will have to apply knowledge and demon-
> strate insights in a public form, they assume a more active stance vis-à-vis
> material, seeking to exercise their "performance muscles" whenever possible.
> (pp. 160-161)

From the standpoint of MI theory, asking students to present their understanding promotes both personalized learning and understanding in depth. (A common-sense view
of teaching and learning is that in order to convey a concept or a body of information
effectively to others, one must thoroughly understand it first.) From the standpoint of
CALL, performances of this kind, supported by such presentation tools as PowerPoint
and HyperStudio, allow students to express their understanding through words, pictures, and sound in a truly harmonious convergence of CALL and MI theory.

The "high end" of CALL, as seen in multimedia presentations based on research,
would be daunting even for native speakers, and therefore poses added stress for
second-language learners; nonetheless, it is a challenge that will stretch learners of
languages or any other discipline in ways likely to be beneficial. Ultimately, working
in the high end of CALL, mediated by MI theory, puts students in the forefront of the
new literacy of our time, or what McAdoo (2000) describes as "exploratory, 'nonlinear' constructivist approaches that students bring to web surfing and computer use"
together with "non-computer skills – the ability to synthesize information, ask further questions, compare, contrast, write, and evaluate" (p. 149). From the perspective
of McAdoo and others, the demands of twenty-first century literacy require more
than "the simple ability to show up for work and do what you're told"; one must be
"able to think carefully and creatively and have interactive skills . . . such as communication, teamwork, and model-based problem solving" (pp. 149-150).

6. Conclusion

Regardless of the goals for students – whether to acquire twenty-first century literacy
(as McAdoo and Dede counsel) or to become effective presenters of their own understanding of various issues in the arts and sciences (as Gardner recommends) – multiple intelligence theory and CALL furnish learners with powerful tools for research
and personal expression. They can facilitate the exploration of the universe of information, the development of understanding about that information, and the presentation of that understanding to others. All that is required is that teachers inspire their
students to aim high in the pursuit of information, knowledge, and insights; teachers
should then step aside and allow students to present their understanding, making the
fullest use of the rich resources that MI theory and CALL have to offer.

References

The American School Board Journal. (Aug. 2001).

Arenson, K. (2001, August 5). Reading statistical tea leaves. *New York Times, College Times.*

Armstrong, T. (1994). *Multiple intelligences in the classroom.* Alexandria, VA: Association for Supervision and Curriculum Development.

de Bary, W. T. (1997). Confucian education in premodern East Asia. In W.-m. Tu (Ed.), *Confucian traditions in East Asian modernity: Moral education and economic culture in Japan and the four mini-dragons* (pp. 21-37). Cambridge, MA & London: Harvard University Press.

Dede, C. (2000). A new century demands new ways of learning. In D. Gordon (Ed.), *The digital classroom: How technology is changing the way we teach and learn* (pp. 171-174). Cambridge, MA: Harvard Education Letter.

Dede, C. (2001). Anytime, anyWear: A forecast of emerging media for distributed learning. [lecture]. At *Syllabus 2001: New dimensions in education technology.* [Conference, July 20-24]. Santa Clara, CA: Syllabus: 101 Communications.

Dryden, L. (1998). What does multiple intelligence theory tell us about CALL? In P. N. D. Lewis (Ed.), *Teachers, learners, and computers: Exploring relationships in CALL* (pp. 95-102). Nagoya: Chubu Nihon Kyouiku Bunkakai.

Dryden, L. (2000). CALL & educational reform in East Asia, mediated by Western and Eastern theories of learning. In P. N. D. Lewis (Ed.), *Calling Asia: The proceedings of the 4th annual JALT CALL SIG conference Kyoto, Japan, May 1999* (pp. 111-114). Nagoya: Chubu Nihon Kyouiku Bunkakai.

Dryden, L., & Morrone, M. (1999). Language learning and self-discovery through multiple intelligence theory. In Korea TESOL (Ed.), *Proceedings of the 1998 Korea TESOL conference* (pp. 27-34). Seoul: Korea Teachers of English to Speakers of Other Languages.

Gardner, H. (1983, 1985). *Frames of mind: The theory of multiple intelligences.* New York: Basic Books.

Gardner, H. (1993). *Multiple intelligences: The theory in practice.* New York: Basic Books.

Gardner, H. (1999). *Intelligence reframed: Multiple intelligences for the 21st century.* New York: Basic Books.

Gardner, H. (2000). Can technology exploit our many ways of knowing? In D. Gordon (Ed.), *The digital classroom: How technology is changing the way we teach and learn* (pp. 32-35). Cambridge, MA: Harvard Education Letter.

Kelly, C., & Kelly, L. (2002). *Internet TESL Journal.* Available from www.iteslj.org

McAdoo, M. (2000). The real digital divide: Quality, not quantity. In D. Gordon (Ed.), *The digital classroom: How technology is changing the way we teach and learn* (pp. 143-151). Cambridge, MA: Harvard Education Letter.

McMurray, D. (1998, August 31). Japan's universities score low on creativity. *The Nikkei Weekly.*

Ministry of Education, *Kyouiku-kate shingikai* (Curriculum Council). (1997, December). *Interim report.* Tokyo: Ministry of Education Press. [English summary of entire book by M. Morrone].

Ministry of Education (2001). *Gakushuu shidoo yoryo (Essentials of educational guidance).* Tokyo: Ministry of Education Press. [English summary of entire book by M. Morrone: For more details, see Ministry of Education curricular guidelines, 2002].

Negroponte, N. (1998). *Being digital.* Boston, MA: MIT Press.

Nord, J. (1997, May 27). *CALL: Basics and beyond.* [Email to the author, cited in Dryden (1998)].

Robb, T. (2002). *The international student discussion lists.* Retrieved from www.kyoto-su.ac.jp/~trobb/#sl

Sperling, D. (2002). *Dave's ESL cafe on the web.* Available at www.eslcafe.com

Thoreau, H. (1854). *Walden.* Boston, MA.

Thornburg, D. (1993). Gardner's theory. In *Challenging times: Working together. Proceedings of computer-using educators conference. Spring 1993.* [CD-ROM]. Santa Clara, CA: CUE.

6

The Terraced Labyrinth of Language Learning

Charles Adamson
Miyagi University
Stephen Shucart
Akita Prefectural University

1. Introduction

1.1 The use of theories of language learning in CALL

Language learning in Japan, as in much of the world, is presently largely not theory-driven in the sense that few of its models are based on a coherent theory of language learning. A quick look at tables of contents of handbooks for the annual conference of the Japan Association for Language Teaching CALL Special Interest Group (JALT CALL SIG), or its proceedings (Lewis & Shiozawa, 1997; Lewis, 1998, 2000), shows that almost no one has specifically presented a model of language learning with the exception of the authors of this paper. Even papers on theory present few models, concentrating instead on experimentation or discussion of points at far higher levels. In EFL in Japan, few teachers seem to base their teaching on explicit models, opting instead for what is commonly known as the *eclectic approach* This paper follows a different path.

The first section presents some essential ideas and describes a new model of language learning, the *terraced labyrinth model* (TLM), which describes important aspects of language learning, acquisition, and emergence at both individual and class levels. The second section relates this model to CALL programming.

1.2 Approaches to model building

Before beginning a discussion of the model, it may be helpful to investigate the various approaches that can be applied to model building in order to determine which one is most appropriate.

1.2.1 Eclectic approach
The eclectic approach to modelling is similar to the eclectic approach in EFL. The teacher or CALL designer applies the criterion of accepting that which works and rejecting that which does not. The problem with this approach is that behind the intuitive decisions about *what works* must be an implicit theory or theories. Otherwise, it would not be possible to make consistent evaluations. These theories are individual in the sense that each person will have a different implicit model based on their experiences and beliefs, so one person's model may be completely different from the next. The models cannot be compared and evaluated because they are not explicit.

1.2.2 Scientific approach
The scientific approach is explicit model building, grounded in experimental fact. Its goal is explaining how and why a phenomenon works. Theories or models are developed on the basis of theoretical ideas and experimental results. Once developed, they arc tcstcd against reality and retained only if they pass the test. Models that fail are modified and retested, or discarded immediately.

1.2.3 Engineering approach
The engineering approach asks how we can use a specific phenomenon to better the quality of human life. It is the practice of turning models into applications that have practical use. The paper on which the TLM is based (Crutchfield & van Nimwegen, 1999) applies an engineering-like approach to the problem of evolution. This chapter retains that engineering strategy and builds on it to produce a complexity-based, evolutionary, terraced model of language.

1.3 Emergence, learning, and acquisition
Gaining a language can be considered from three different perspectives. One is *learning*, in which specific elements of language are explicitly taught and practiced. A second perspective is *acquisition*, in which a person gains language through exposure to meaningful input. Finally, in a complexity-based model, language is said to *emerge* from the interaction between the variables involved. These three together are referred to as emergence, learning, and acquisition (ELA).

While it is occasionally necessary to specify the processes of gaining a language, the differences between these three terms are frequently of little importance. Thus, in keeping with general practice, the term learning will be used here to encompass all three processes, when precision is not necessary. Context will identify any contrastive use of this term.

2. Characteristics of language

Having clarified the modelling approach, the various aspects of language involved in the model will be discussed briefly.

2.1 Language as a hierarchy

The concept of language as a hierarchy implies that no matter which language feature is considered, there is a set of prerequisites that must be present before that feature can be mastered by the student. For example, before a vocabulary item can be mastered for oral production, the student must have control of the phonetic values present, an implicit knowledge of the parts of speech, and the ability to link from the verbal representation to meaning, to list only a few necessary elements.

Although it would take a huge effort, it is theoretically possible to specify the prerequisites for all language features. The most efficient and natural form for this specification is a tree with the highest level processes as the outer leaves and the physical processes as the roots.

Once the tree is defined, the prerequisites for any language feature would be those nodes that form a path leading from the base to that feature. The different levels can be defined by the number of nodes between the feature in question and the base of the tree. The language labyrinth is simply a model of the language based on its hierarchy.

It should be noted that this tree is not the simple one used in most linguistic theory, but is extremely large and complex, since there are many possible paths and sequences to language learning. The language labyrinth is constructed by allowing the connections between nodes to cross or by repeating nodes at more than one location.

2.2 The classes of variables that influence language usage

It is obvious that factors other than the language itself affect the efficiency of usage, both in and out of the classroom. Most people have experienced the typical ups and downs that seem to make using the language either extremely easy or virtually impossible. Factors ranging from mood to the temperature of the location to the methods used by the teacher when the speaker studied the language have wide-ranging effects on how well the language is used. The factors involved are almost infinite in number, but can be organized into seven classes of variables, which may or may not be present in any given situation. These factors apply both in the classroom while the student is studying and outside during real communicative use of the language. The effects are cumulative over time with the situation in the classroom affecting later usage.

2.2.1 Student (or speaker) variables

These are the factors that come from within the individual speaker and include, but are not limited to, such things as the mood, linguistic ability, prior learning and experiences, audio discrimination, motivation, health, and expectation. Another important aspect is the amount of exposure and practice the student has had with the language.

2.2.2 Peer variables

This class of variables concerns the effects of interrelationships between the various members involved in communication. When a single supportive group forms, usage is enhanced, but if the group contains cliques and factions, performance can be severely inhibited. Also included here are the effects of distributions of the variables across the group. For example, a group in which the individuals exhibit a broad range of motivations and linguistic ability is likely to differ substantially from a group in which the distribution of these variables is more restricted.

2.2.3 Teacher variables
This class of variables relates to the skills, knowledge, training, and experience of the teacher as well as such factors as personality, current mood, and health.

2.2.4 Method variables
These variables concern the specific method, or lack of method, that a teacher uses, and clearly the choice of method effects the outcome. Simply naming the methodology (e.g., the silent way or audiolingualism) is insufficient, since each teacher claiming to use a specific method will interpret it in a complex and differing way. Therefore, *method* must refer to the actual classroom practices rather than just being a name.

2.2.5 Material variables
The physical materials employed for both language study and everyday use can have an effect. Material-related variables include specific content, organization, and layout as well as the media employed and the professionalism and difficulty of the material. The presence or absence of computers, books, audiovisual equipment, and the like may have important effects.

2.2.6 Environmental variables
The characteristics of the physical location affect language usage. These factors include temperature, comfort of the chairs, colour of the room, and amount of free space. In Japan, particularly in older public universities, this also includes the overcrowded and dirty conditions often found; air conditioning and heating are also likely important factors.

2.2.7 Administrative variables
Administration refers to factors such as scheduling, curriculum, availability of equipment, and social expectation. They relate to the school and society but are beyond the control of the student or teacher. Included here are situations like those in Japan where college entrance exams exert a strong influence on the content of school curricula, in turn diverting student attention from the communicative use of language to an overriding concern with exam scores.

2.3 Fitness landscapes
Fitness landscape is the term applied by evolutionary specialists to a multidimensional graph on which the dependent variable (fitness) is the ability of a population to survive and reproduce and the independent variables are those factors that influence this ability. Each combination of independent variable values will have a fitness associated with it and a plot of all these values will form a surface, called a landscape because it has flat areas, valleys, and peaks. A single point on the landscape can be represented by a large binary number; each axis is represented in a segment of the number that contains one bit for each unit on the axis. For example, if the variable has three possible values plus zero, the segment must contain three bits, and the possible settings are 000, 001, 010, and 100. Only one bit can be set to 1 because the variable has only a single value at any given time.

Complexity specialists have adopted the term *fitness landscapes* and now use it to describe almost any multidimensional graph with a single dependent variable. The model developed here incorporates three different fitness landscapes.

2.3.1 Characteristics of fitness landscapes

A landscape with a small number of variables will have a single peak, but as the number of variables increases, the number of peaks increases, until the entire landscape is covered with steep-sided peaks and valleys. Landscapes may be dynamic in that the fitness of a particular point may change over time (Kauffman, 1993, 1995).

2.3.2 Kauffman's NK landscapes

Stuart Kauffman of the Santa Fe Institute has done extensive research on landscapes where the variables interact. He refers to these as NK landscapes because there are N variables that each interact with K other variables. He has found that, among other features, these landscapes are extremely rugged and small changes in the variable values usually causes extremely large changes in the fitness (Kauffman, 1993, 1995). Although not as well investigated, landscapes where K varies for each N appear to be even more rugged and unpredictable. Language learning matches this last example well, which may explain the frequently conflicting results obtained when repeating learning experiments. Because of the ruggedness, extremely small, unreported changes in initial conditions may move students from a high peak to a valley.

3. The terraced labyrinth model

The original TLM (Crutchfield & van Nimwegen, 1999) is a tree-based model of punctuated evolution, where the fitness represents the probability of reproduction. Leaves represent species, and branches are paths to the emergence of new species. There are no restrictions on the number of branches radiating from a single leaf, although each branch may connect to only one other leaf. The model includes the basic components listed below.

3.1 Genotypes

A *genotype* is a very large but finite binary number in which each bit represents the presence or absence of a specific feature. The instantiation of a genotype is called a *phenotype*.

3.2 Subbasins

A *subbasin*, termed a leaf in the above sections, is a multidimensional volume where the independent variables are only those factors that affect the reproductive ability of the species represented by the subbasin. Each subbasin will have its own set of pertinent variables, which may or may not affect other subbasins. Variables that do not affect the subbasin in question are called *neutral variables*.

3.3 Portals

A portal is a subset of bits, called a *constellation*; when set to a specific value, it will allow movement to the next higher subbasin. This movement represents the emergence of a new species. It should be mentioned that a description of the new feature cannot be predicted before its initial emergence. However, when a reproducing pair of individuals has moved through a portal, the population will rapidly shift to this new level.

3.4 Fitness
Since the original terraced labyrinth is a model of evolution, its fitness measures the probability of an individual surviving and having children.

4. A TLM for language learning

This section introduces a modified TLM that describes the entire learning process in a concise, practical fashion. The model is comprehensive in the sense that it includes language emergence, acquisition, and learning.

4.1 Learning a specific linguistic feature
Although there are many ways of analysing the act of learning a single linguistic feature, one approach is to develop a model where the features arise from a set of subskills that must be learned before the feature in question and a set of variables that affect the efficiency of the feature's usage. For example, before being able to use a word correctly in speech, a student needs at least minimal control of meaning, pronunciation, stress pattern (including the influence of sentence stress), and knowledge of contexts appropriate for the word,

In other words, there is a set of prerequisites for the process and the effectiveness of the student's usage is influenced by a large set of variables. Additionally each of the prerequisite processes can be modelled in a similar way, as can the prerequisites for the prerequisites – all the way back to the person's DNA.

4.2. A TLM for language learning
The following outlines a modified TLM that applies to an individual's physical and mental characteristics, rather than to the population characteristics described in the original model, and which specifically refers to a language labyrinth, the portion of an individual's terraced labyrinth relevant to languages.

4.2.1 New definitions of fitness
In the model, landscapes arise in three places, each with its own definition of fitness. The first is the overall language labyrinth in which the subbasins are sequenced. The second models the process of searching for a portal within a subbasin. The third shows the relation between actual usage of the feature that has emerged and ELA variables.

4.2.1.1 Fitness between subbasins in the language labyrinth
The first fitness value is a sequencing variable that appears on the y-axis of the tree diagram. The x-axis contains each possible phenotype, but because of the near infinite number of potential phenotypes, only those connected by a portal to the next level are actually shown. This fitness shows the order in which the features emerge. For example, if a feature has a fitness of 1,147, this means that 1,146 other features must emerge sequentially before this particular feature can emerge. A complete TLM

of a human would place the DNA sequence at level 0. Eventually, at some fairly high level, there would be a branch that was clearly related to language. It is this branch that is of relevance here.

The only way to advance to the next higher level is to complete a constellation for one of the portals leaving the current subbasin. Once this is accomplished, the new feature emerges. Although emergence takes a certain amount of time, for the purposes of this paper, this factor can be ignored.

4.2.1.2 Fitness for a genotype bit within a subbasin

Whether or not the characteristic described by a single genotype bit will actualise is dependent on both the states of the other bits and the environment. Therefore, each bit will have a fitness landscape associated with it. The fitness will be the probability that the characteristic will change and the independent variables will be the states of the other variables and whatever features of the environment are experimentally found to be pertinent. At present, this is believed to be limited to meaningful linguistic input for features in the language labyrinth. All other variables would be neutral.

4.2.1.3 Fitness for use of a feature after emergence

This is a measure of the probability of appropriate and accurate use of a feature once it has emerged. The independent variables are contained in the ELA classes described above.

Landscapes where fitness measures ease of language use are of the NK type, since many variables interact with each other. For example, if the student is in a bad mood, this may affect the teacher's mood, and the combined effect of moods plus the temperature would be different from the individual effects of each isolated factor. It is possible that many ELA variables will be neutral with regard to any given feature.

4.2.2 Specification of the genotype

As in the general model, the genotype is a large, but finite, binary number in which each bit represents the presence or absence of specific subbasins. Subbasins can always be replaced by their prerequisites. Here, only those subbasins that are pertinent to language are included in the model.

4.2.3 Subbasins as individual aspects of language

A subbasin represents different things at different levels. At lower levels, a subbasin would represent something as specific as possessing the particular muscles required to make a particular sound or move the eyes across letters on a page. At a slightly higher level, the subbasin will represent the ability to control those muscles, with their presence a prerequisite. At a still higher level, another subbasin might represent the ability to incorporate surrounding sounds in the phonetic characteristics of the particular utterance or the letters into words. At still higher levels, subbasins would represent larger and more complex features of the language, including grammar, vocabulary, and function.

4.2.4 Portals between linguistic subbasins

Once all the prerequisite bit settings have been acquired, the new feature will be available. The feature will be internally present but whether it is used correctly in communication will depend on the ELA variables. Practice is one aspect of these variables and it becomes particularly important in relation to gaining control over the feature once it has emerged.

4.2.5 Effects of input
Language cannot be learned without linguistic input. The TLM predicts that the genotype settings will change according to the linguistic input received and the concurrent settings of the other bits in the genotype. Future research may show that the ELA variables also have an effect, but until such evidence is available, Occam's razor must be applied and the ELA variables considered to influence only the usage of a feature.

At present it is impossible to determine precisely the students' location in the language labyrinth, and hence what input they need next. In addition, students in a class will all be at different levels. The best temporary solution may be to provide students with a large amount of comprehensible input. As details of the tree develop, it should be possible to tailor the input more closely to the needs of the next subbasins, but at present comprehensibility seems to be the best route. This will gradually change as the details of the model are discovered and CALL programs become available.

4.3 Differences between learning and acquisition
Over the years there has been much discussion of the differences between learning and acquisition. The TLM helps clarify the differences, and the acquisition/learning dichotomy may turn out to be inappropriate, at least at a deep level.

4.3.1 Language acquisition
Acquisition, according to this model, is the adjustment of the student's genotype settings through exposure to comprehensible target language input. The settings change in a semi-random way that depends on the statistical features of the language input. Since the individual's path up the language labyrinth will be idiosyncratic but linear, acquisition turns out to be both linear and nonlinear: The individual path for acquired language is linear, but the paths for a group of students and the potential paths for future acquisition are nonlinear. Acquisition is generally non-conscious, although the student may be very conscious of meaning and a limited number of language features.

4.3.2 Language learning
Learning (or being taught) a specific feature is usually an attempt to jump (or be pushed) out of the tree. A feature is presented to the student whether or not the prerequisites are present and the student is expected to learn the feature, or fail, the latter bringing with it many negative consequences. The students will attempt to place the new feature into their general tree. Some will find the prerequisites available and the feature will become part of their language labyrinth. Others will find a place for the feature in another part of the general tree. These students may consciously understand the feature but not be able to employ it as fluent language. The remainder of the students will find they are missing prerequisites and will neither understand nor be able to use the new feature. If all prerequisites are in place, learning and acquisition are similar and the relative effectiveness from the teacher's point of view may depend on content, not method of presentation.

4.4. Possibilities for specifying and verifying the model
Although it will require much work, the model can be specified in detail and verified. The first step would be to select specific language features and determine the necessary prerequisites. The prerequisites would then be analysed for their prerequisites. This process would continue until a level was reached where one could assume that students would bring the features with them when beginning study of the language. At the same time, the initial features could be used as a starting point for determining prerequisites for upper level features. Eventually the entire language labyrinth for a given language could be specified in detail. Throughout the process, concurrent research using MRIs, cognitive psychology, traditional linguistics, and other methods could be used for additional data and verification of the subbasins at each level. As advances are made in medical science, it may become possible to compare the language labyrinth to events in the brain itself.

5. Application of the terraced labyrinth to CALL programming

As mentioned earlier, most recent work in CALL has been done without the advantages of an explicit working model of language learning. The terraced labyrinth gives us a potential model. The remainder of this paper explores possibilities for developing CALL programs based on this model. This exploration will be a bottom-up process, beginning at the lowest operational level, the TOTE, and building up to an entire program from there.

5.1 The TOTE and the terraced labyrinth CALL model
5.1.1 The TOTE model
Test-Operate-Test-Exit (TOTE) was proposed by Miller, Galanter, and Pribram (1960), who recommended it replace stimulus-response as the basic unit of behaviour. Each TOTE unit has a goal, a target or condition it attempts to achieve. The first step in the process of applying it is to test if the goal has already been achieved. If so, the unit exits and the process ends. If the goal has not been achieved, the unit operates in some way and then tests again. This cycle of *operate and test* continues until the goal is achieved and the unit exits.

The classic example of a TOTE is the process of hammering a nail. One first tests to determine if the nail is level with the surface by looking or feeling. If it is level, one exits. If not, one operates by hitting the nail with the hammer, and then tests again. If the nail is now level with the surface, exit. If not, operate and test again. The operate (hammer) and test (looking or feeling) cycle continues until the goal (nail level with surface) is confirmed, at which time one exits.

Miller et al. also propose that real cognitive processes are the result of hierarchies of TOTEs. This means that the exit process will pass the overall process to another TOTE at either a higher or lower level in a large system of TOTES. These systems are reminiscent of the TLM, which has subbasins organized into a hierarchy.

5.1.1.1 The terraced labyrinth TOTE
In the terrace labyrinth CALL model, a TOTE can be used in conjunction with each subbasin. The goal is simply to determine if the student has learned the linguistic

feature represented by the subbasin. The specific subbasin is determined by overall program goals. However, the process is more complex than hammering a nail.

The test process, which begins the TOTE process, is more complex than the simple question of whether or not the student has learned the process. From the TLM, it is clear that there are three possible results. First, (1) the student may know and use the linguistic feature. However, when turning to the possibility that the student does not know the process, there are two distinguishable reasons: (2) the feature may not have emerged yet, or (3) the feature has emerged (all the prerequisites are present) but is not being used accurately.

In the case of (1), the TOTE simply exits.

If the test results in (2), the operation will provide instruction and practice with the feature. However, before doing this, the operation will check for the presence of prerequisites. If one or more prerequisite is not present or used, the operation will initiate new TOTEs at that level. This may be repeated for the prerequisites of the prerequisites and so on, creating a tree of TOTEs that continues to expand until every branch reaches a level where the test result is (1) or (3).

For (3), the operation will teach the student the feature. In most cases, this instruction will consist of the presentation of comprehensible input that draws the student's attention to the feature as meaning. This may be done implicitly or explicitly, depending on the feature. Adamson (1999) offers some guidance by comparing Reber's implicit approach to acquisition with Anderson's ACT*, an explicit teaching procedure. Adamson finds that complex knowledge is best presented by implicit methods while simple procedures are more effectively presented through explicit methods. However, there are exceptions when the student has previously received explicit instruction. With the TLM, it is not yet clear which approach is appropriate in any given case. However, it is probably true that at the very lowest levels, the processes are well out of the realms of consciousness, thus suggesting implicit methods be used. At higher, more clearly linguistic levels, however, it is not obvious which method would be more effective. Once the TLM is specified in detail, this question can be answered experimentally. Until that time, the CALL program author can offer the student a choice, providing both explicit and implicit instruction where possible. The CALL program can retain data on student responses and gradually build up a data bank that will allow an informed decision to be made.

For the situation where the ELA variables are holding the student at a low point on the usage fitness landscape, a psychological approach is needed. For the foreseeable future, incorporating such materials in CALL programs will probably be only partially successful. Person-to-person contact is likely to be more effective. However, additional practice and some support are possible. Each year, increasing numbers of psychological aspects could be incorporated into CALL programs, partly due to the increasing power of computers, still expanding extremely rapidly.

5.2 General description of a terraced labyrinth CALL program

It is now possible to describe a complete terraced labyrinth CALL program. Each subbasin of the TLM will have an associated module in the program. The module will contain the TOTE process and links to the TOTEs for subbasins at immediately higher and lower levels in the hierarchy.

Both tests and operations could be multimedia-based, incorporating graphics, video, audio, and the written word presented in various combinations. For examples of similar general CALL material developed in Japan, see Lewis (1998, 2000), Lewis and Shiozawa (1997), Ryan (2000), or Kruse and Peterson (in press). The materials incorporated in the program will depend on a number of factors, some of the more important being course goals, the knowledge and experience of materials writers, the creativity of all concerned, and probably most importantly, the budget.

5.3 Data collection
The program could contain a data analysis module that would provide the teacher with feedback about individual students and the program designer with data that would become the basis for modifying or totally redesigning individual TOTE modules.

5.3.1 Data for the program designer
The program designer would need to know whether correct answers to the test items accurately indicate the emergence of a particular feature. The designer is interested in collective data about all students and particularly whether decisions at one level are accurately predicting student behaviour at both higher and lower levels in the hierarchy. Standard statistics and comparative data across levels will be of use here.

The designer would also be interested in the relative difficulty of test items and the correlation between item difficulty and the presence or absence of the feature, information provided by Rasch analysis (Henning, 1987[1]). Rasch analysis is useful because it allows the difficulty of an item to be determined absolutely, rather than relative to the population of examinees. Student ability can also be determined likewise, without regard to the items themselves. Rasch analysis would be useful since, although in theory each item should clearly indicate the presence of the feature, in reality the situation is usually fuzzy. By knowing the comparative difficulties of items, any known weak ones could be removed and replaced by more efficient examples.

5.3.2 Data for the teacher
The teacher would be interested in a different set of data. Probably the most important statistics for the teacher will be the amount of student time spent using the CALL program while working on each TOTE. This time will correlate with learning, since student study time is being efficiently used for either testing or learning at the appropriate level, eliminating much inefficiency. Large gaps would indicate that the student had lost focus and was not concentrating on the program. Such a situation might well call for psychological intervention by the teacher. Other data of interest might be the frequency of answering questions or responding to learning situations. The inclusion of a periodic assessment module with questions explicitly above and below the student's current subbasin would allow both teacher and student to accurately judge student progress. Such data could eventually allow accurate teacher estimates of the time required for specific students to reach specified levels in the hierarchy. If study on the CALL program is being supplemented by classroom work, it would be important for the teacher to know each student's current level in order to maximise the effectiveness of classwork.

5.4 Actualizing a terraced labyrinth CALL program
Although the full terraced labyrinth is huge, and designing a complete CALL pro-

gram based on it would be a daunting task, it may be possible to build the full program piece by piece. A CALL designer could begin by selecting a single subbasin from the course goals and programming the TOTE for it and the subbasins in the layer at the level immediately below. For this first module, it would be assumed that any missing subbasins from even lower levels would emerge from comprehensible input presented in operation processes. This same assumption is also made in many other CALL programs so the results of using this module alone would be no worse than other approaches. The next step would be to select another subbasin from the course goals and develop a module for that and the next lower subbasins. This would continue until all course goals were included in a module. As each module was programmed, the designer would check the previous one for overlap, the appearance of a single subbasin in multiple modules. When overlap occurred, the designer would, rather than prepare a new module, use the one already in existence. This would reduce work and allow the modules to be linked together into a tree, corresponding to the terraced labyrinth. Once all course goals were accounted for, the designer would start working back through subbasins at lower levels, preparing a module for each. Eventually the entire tree would be available. An advantage of this approach is that as each module was written, it could come online and be used profitably with students.

A rule-of-thumb which many teachers have found useful claims a student should get between 70% and 80% of responses correct. Any more and the work is too easy. Any fewer and it is too hard. In either case, boredom readily sets in and the student may give up or become distracted. This factor could be built into the program by the occasional inclusion of easy work based on the next higher subbasin or particularly difficult work at the current level. Also, easy review work for previous TOTEs would allow students to understand how much they have progressed. The materials used could be selected on the basis of the Rasch analysis discussed above.

Through such sequential design and programming, a CALL program to move the student to a specific point in the terraced labyrinth could eventually become a complete package. Each development phase after that would include the initial program as a subsection and take the student to ever higher levels.

5.5 Using a terraced labyrinth CALL program

From the student's point of view, the terraced labyrinth CALL program would provide a unique experience. The program would be able to adapt to the student since, even if the program started off at too difficult a level, it would quickly adjust itself The student would also have the psychological advantage of knowing that he or she was progressing through the material because an assessment module would show that, after the drop following the initial testing and placement in the appropriate module, steady progress was being made. This progress supplemented by preview and review materials should make the CALL package very attractive for the student.

6. Conclusion

The terraced labyrinth is a new model of language learning, based on advances being made in the understanding of complexity. It is a tree-like structure through which the student progresses on a linear path, the tree trunk. However, no two students would be likely to follow the same path and the possibilities for additional learning are always nonlinear and complex.

A CALL program based on the TLM could be constructed in stages or modules. Each of these stages could be used with students without waiting for the complete program to be assembled. As the stages were completed they would, where appropriate, be linked together so that eventually a complete program would become available.

This discussion of the TLM has been entirely theoretical. It is, however, relatively easy to demonstrate the validity of the model through further analysis and experimentation. It is recommended that such research be conducted and that work be started on a terraced labyrinth CALL program.

Notes

1. Henning (1987) contains a serious typographical error. Formula 8.10 on page 123 should read (d-b) rather than (b-d) as written.

Acknowledgements

The authors would like to thank Charles Nelson, University of Texas at Austin, for comments on an earlier version of this paper, the preparation of which was partially funded by a research grant from Miyagi University.

References

Adamson, C. (1999). Language learning – implicit and explicit? *Miyagi University Nursing Review, 2*(1), 106-113.

Crutchfield, J. P., & van Nimwegen, E. (1999, February). *The evolutionary unfolding of complexity.* Santa Fe Institute Working Paper 99-02-015.

Henning, G. (1987). *A guide to language testing.* New York: Newbury House.

Kauffman, S. (1993). *The origins of order: Self-organization and selection in evolution.* Oxford: OUP.

Kauffman, S. (1995). *At home in the universe: The search for the laws of self-organization and complexity.* Oxford: OUP.

Kruse, M., & Peterson, S. (Eds.). (in press). *JALTCALL2001 Proceedings.* Nagoya: JALT CALL SIG.

Lewis, P. N. D. (Ed.). (1998). *Teachers, learners, and computers: Exploring relationships in CALL.* Nagoya: Chubu Nihon Kyouiku Bunkakai.

Lewis, P. N. D. (Ed.). (2000). *Calling Asia.* Nagoya: Chubu Nihon Kyouiku Bunkakai.

Lewis, P. N. D., & Shiozawa, T. (Eds.). (1997). *CALL: Basics and beyond.* Nagoya: Chubu Nihon Kyouiku Bunkakai.

Miller, G. A., Galanter, E., & Pribram, K. H. (1960). *Plans and the structure of behavior.* New York: Holt, Rinehart & Winston.

Ryan, K. (Ed.). (2000). *Recipes for wired teachers.* Nagoya: JALT CALL SIG.

Appendix A

The terraced labyrinth in graphical form.

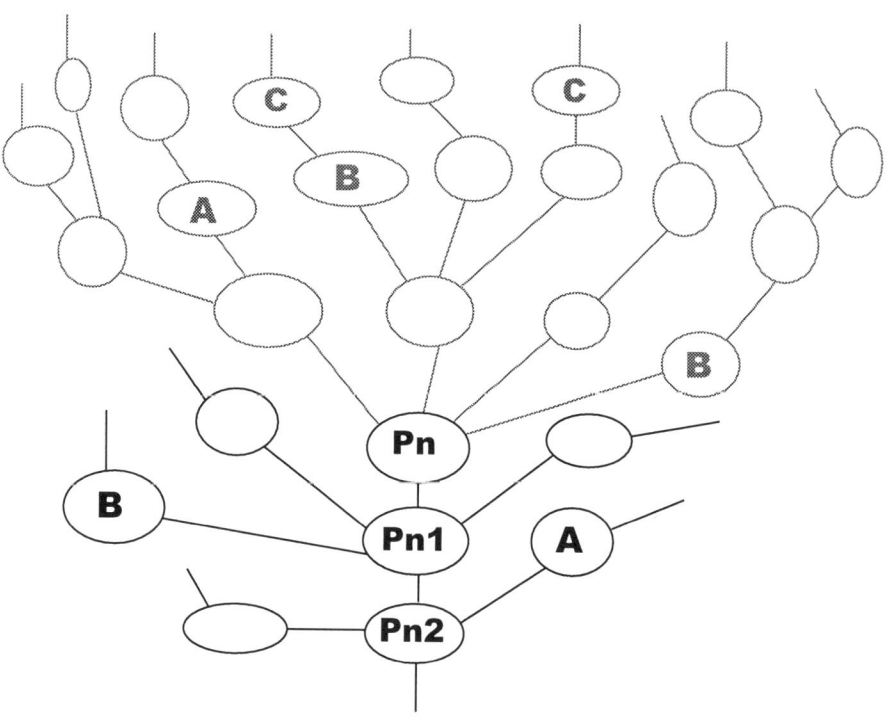

Figure 1. A section of the terraced labyrinth.

This diagram represents a small portion of the overall labyrinth for a language. The leaves (circles) represent individual linguistic features or skills, which at low levels might be the ability to control certain muscles, at an intermediate level might represent control of a single word or grammatical functions such as the plural, and at the very highest levels represent functions, factors related to dialect and register, and artistic use of the language. In the diagram, the sequence of circles labelled Pn, Pn-1, and Pn-2 represent the top of the line of features that have already been acquired. The green portions of the labyrinth represent potential futures paths of acquisition and the red portions represent branches that are no longer accessible.

Features may appear in multiple locations in the labyrinth as shown by the leaves labelled A, B, and C. The leaves labelled A and B show that features that are in inaccessible branches will also appear in the still acquirable portions of the labyrinth. The demonstrates the a single feature will appear in more than one path. The is no limit to the number of times that a single feature may appear in the labyrinth.

Section 2:

CALL Resources

7

Corpus Linguistics in Japan: Its Status and Role in Language Education

Monika Szirmai
Hiroshima International University

1. Introduction

The Japan Association for English Corpus Studies (JAECS) was established in 1993, and Takefuta published the results of his corpus-related research in Japan as early as 1981. So, in these respects corpus linguistics has a relatively long history in Japan. On the other hand, even five or six years ago, when mentioned to language teaching colleagues in Japan, it was usually greeted with puzzlement or embarrassment. Behind these responses was ignorance about what corpus linguistics meant and also perhaps the suspicion it was just one more obscure area of research irrelevant to the practical needs of language teaching.

Nowadays, largely due to its being a selling point for EFL dictionaries, corpus linguistics has become a kind of buzzword. Despite this, what exactly it means may still not be clear for everybody – as is very often the case with buzzwords. That may also help explain why some CALL practitioners in Japan do not yet appreciate the relevance of corpus linguistics to CALL. At the same time, though, for increasing numbers of language teachers in Japan, as elsewhere, the use of corpora for CALL is an exciting and effective element in their language classes.

2. Corpus Linguistics and CALL

2.1 Defining Corpus Linguistics

To help demonstrate that it falls within the scope of CALL, let us clarify terms. Corpus linguistics is an interdisciplinary field closely related to or overlapping computational linguistics. It is the study of electronic corpora, which Sinclair (2001) defines not simply as a random collection of texts but one "gathered on the basis of external criteria (Clear, 1992) . . . and the claim is implicitly made that an investigation into the internal patterns of the language used will be fruitful and linguistically illuminating."

As many research tools used in these two fields are identical, the difference between computational and corpus linguistics is not easy to tell, lying basically in approach and research goals. Roland Housser (1999, p. 1) says: "The goal of computational linguistics is to reproduce the natural transmission of information . . . on a suitable type of computer. This amounts to the construction of autonomous cognitive machines (robots) which can communicate freely in natural language." Housser (1999, p. 13) adds that, in order to achieve this goal, computational linguistics must first find the "solutions to the most basic tasks of natural language analysis."

The goals of corpus linguistics, as Kennedy (1998, p. 1) writes, are both wider than this and also more pedagogically oriented: "Corpus linguistics is not an end in itself but is one source of evidence for improving descriptions of the structure and use of languages, and for various applications, including the processing of natural language by machine and understanding how to learn or teach a language." To paraphrase, the goal of corpus linguistics could be defined as the empirical study of language based on electronic corpora, carried out with the help of computer software, with the aim of using its results in a range of fields, such as computational linguistics or language teaching. Kennedy (p. 2) adds "Work relevant for corpus linguistics is being done in many fields, including computer science and artificial intelligence, as well as in various branches of descriptive and applied linguistics."

It should also be mentioned here that there is confusion about what constitutes corpus linguistics. On the one hand, the term is widely used in a way implying a field of study similar to the terms computational, theoretical, or applied linguistics. On the other hand, it is often labelled as "just" a methodology. The same paper may refer to it as an academic field and a few pages later as a methodology. The author of this paper considers corpus linguistics as a field of study rather than just a methodology.

2.2 The scope of CALL

Levy mentions that neither the scope of CALL nor its relationship with other related fields are clear (1997, p. 6). He adds that a number of relatively new disciplines have influenced its development, and he devotes a whole section of his book (Chapter 3) to "an interdisciplinary perspective." Although corpus linguistics is not mentioned here, he discusses the results of the influence exercised by computational linguistics (pp. 60-65). As many tools and techniques mentioned by Levy are identical to those used in corpus linguistics, such as concordancing, tagging, and parsing, it would be difficult to exclude certain aspects and practices of corpus linguistics from the scope of CALL. In fact, as language teaching is not among the goals of computational linguistics but of corpus linguistics (see Kennedy's quote above), it seems essential that teachers interested in CALL should get acquainted with it and welcome any possible contribution it offers.

2.3 Corpus linguistics, language teaching, and CALL

Corpus linguistics seems to lend itself to a four-step approach from a CALL practitioner's point of view (Fig. 1). The first three steps cannot be regarded as CALL strictly speaking but are necessary for getting acquainted with the ideas and possibilities of corpus linguistics.

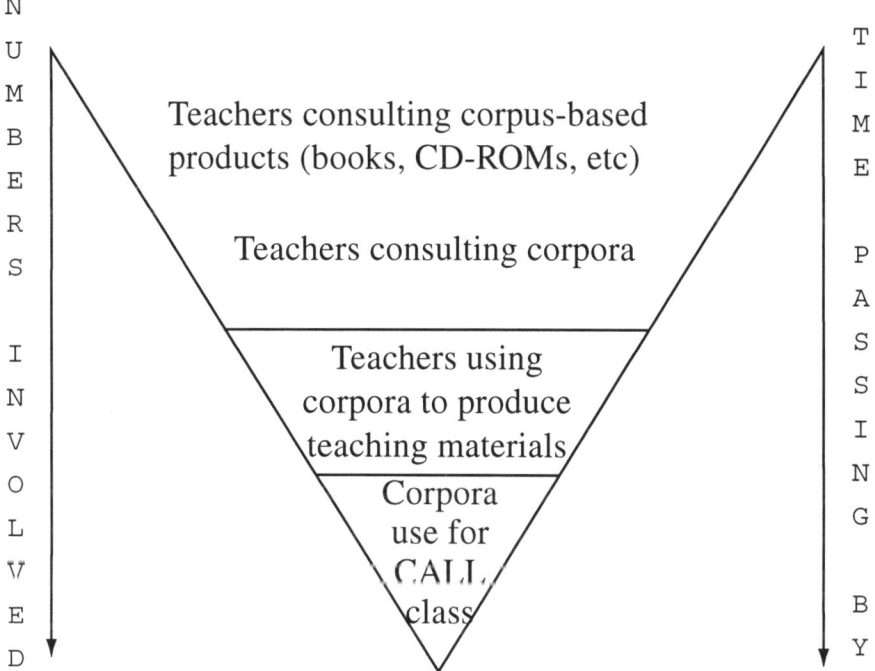

Figure 1. Teachers' involvement in corpus activities leading to corpus-based CALL.

The first stage is where teachers use corpus-based products, such as electronic dictionaries on CD-ROM, word lists, or other printed resources, for consultation. As the second stage, teachers consult a corpus directly using concordancing software. The teacher can look for answers that cannot be found in reference books, because she is not familiar either with a particular structure or use of a word, or with the particular variety of English in question. The third stage is using corpora and corpus tools in creating teaching materials on paper. Finally, the fourth stage can definitely be considered part of CALL, and that is when the students are actually using the corpus and tools the same way as teachers do in the second or even third stage. In addition, basic corpus tools can become hidden, integrated into an interface or application completely different from the simple interface the researcher is used to in the case of corpus tools. For example, the program called GLOSSER strives for "The integration of existing language technology tools: morphological analysis and disambiguator, dictionary and corpus processing" (Roosmaa & Prószéky, 1998, p. 106).

2.4 Some corpus tools

Corpus tools may be less spectacular than other CALL applications, such as multimedia applications, speech recognition, etc., which people often have in mind when talking about CALL. However, they can contribute just as much to the success of learning a language. As there is a vast body of literature describing both technical aspects of corpus tools and their use in detail (see e.g., Biber, Conrad, & Reppen, 1998, McEnery & Wilson, 1996, or Barnbrook, 1996), in this paper there is only a brief mention of these tools.

The most widely known and used tools are concordancers. Their popularity in Japan is probably due to the numerous presentations and demonstrations given mainly by publishers of EFL dictionaries and lexicographers as part of publicity at conferences for language teachers. (Let us not forget that most teachers are *not* researchers interested in teaching *but* teachers interested in research with a desire to apply findings to improve teaching.)

The first explicitly corpus-based dictionary, the *Collins Cobuild English Language Dictionary* (Sinclair, 1987) came out in 1987, and has been followed by many others based on the same principles. Due to the considerable success of these EFL dictionaries, which use corpora and corpus tools for their lexicographic research and compilation, credibility has increased among language teachers of the potentials and merits of using the same methods as lexicographers in their own teaching. Some of the earliest publications about the use of concordancing in language teaching include *Concordances in the Classroom* (Tribble & Jones, 1990), *Should You Be Persuaded: Two Samples of Data-Driven Learning* (Johns, 1991), and *Towards Classroom Concordancing* (Deschamps, 1992). Some of the recent ones in Japan are *Concordances: Production and Use for Teachers and Learners* (Ruthven-Stuart, 2000) and *Corpus tools in Language Teaching and Learning* (Szirmai, 2001).

Tagging, parsing, and lemmatising are not generally used in the classroom unless the aim of the course is grammar study. An interactive tagging program can help students learn parts of speech in a more enjoyable way and with instant feedback. Similarly, identification of the functions of words within the sentence, or of words belonging to the same lemma can be studied with the help of a parser or lemmatiser. It can be used with native speakers learning the grammar of their mother tongue or students studying a foreign language who need such linguistic knowledge.

この数年来、小畠村の閑間重松は姪の矢須子のことで心に負担を感じて来た。数年来でなくて、今後とも云い知れぬ負担を感じなければならないような気持であった。

Figure 2. Japanese text with no word boundaries.

Morphologic analysers should also be mentioned here despite the fact they are not widely used in the teaching of English. However, they can considerably facilitate the study of agglutinating languages, where morphological structures may not easily be spotted. In the case of Japanese, which also belongs to this group of languages, it is even more difficult to decipher morphological information, as word boundaries are not marked in writing, as can be seen in figure 2.

3. The educational climate in Japan

It would be impossible to fully understand the ways and extent that corpus tools can help language learning in Japan without describing briefly the present situation and traditions in education. Recent changes have been considerable, and more are in the pipeline. Conversation classes have already been introduced in high schools, the choice of approved textbooks has increased, and bringing English into primary school is being discussed. However, attitudes do not change overnight with the introduction of new laws or regulations. Let us consider student and teacher attitudes.

3.1 The students: Rote learning
According to the traditional view of Japan, most children go to cram schools from a relatively early age in order to get into the educational establishment of their (or rather their parents') choice. Data available from the Japan Information Network (www.jinjapan.org/stat/stats/16EDUA1.html) proves that this is very much so in the case of junior high school students. They spend most of the day at their regular school and go to cram schools in the evening, often only returning home around 10pm or later. Still, the result is "the shocking discovery that Japan ranks worst in Asia after Laos and Cambodia in English-language ability despite the vast resources and time spent teaching the language" (Clark, 2000). Although it is not clear what data Clark refers to, the *TOEFL Test and Score Data Summary,* available at <www.toefl.org/ pubs/resdloadlib.html#summaries>, seems to support his statement.

Dependence on rote learning alone seems one of the biggest obstacles to foreign language acquisition. As getting into a secondary school or university means sitting for exams, it becomes a question of life and death, sadly sometimes even literally, to pass them. To facilitate preparation for these exams, and possibly try to lower student anxiety level, there are large numbers of word lists and test preparation text available in Japan. These give candidates the reassurance that once they have learnt these lists and grammar points, and are able to regurgitate them all, they cannot fail the exam and will do well, getting into university. Most dictionaries add to this effect by indicating whether a word should be learnt by the end of middle school, high school, or beyond. Examples of such dictionaries are *Genius English-Japanese Dictionary* published by Taishukan Shoten or *Lighthouse English-Japanese Dictionary* by Kenkyusha, which are generally recommended for students. The effect of all this is compounded by the fact that teachers creating entrance exams are also under pressure to respect these lists. All this encourages – and in a way rewards – students who believe that learning languages means learning word lists and grammar rules and being able to identify them on paper when needed. Most of these tests require no creativity or production on the examinee's part; entrance exams are mostly multiple-choice tests and universities use computer readable mark sheets that give instantaneous results.

Another type of rote learning that Japanese children experience from the very beginning of their studies is learning to read and write in Japanese. When I started school, it took me less than one school year to learn all the letters and combinations I ever needed in order to read and write in Hungarian. I only had to practice reading after that, but never again did I have to memorise a new letter or combination. However, in Japan children have to learn about 2,000 *kanji* characters throughout their school years. Some characters consist of many strokes and students have to memorise even the stroke order. This problem is aggravated by the many different readings a character can have. Thus learning to read and write in Japanese, their mother tongue, also means rote learning to a large extent. Why should a foreign language be different then?

3.2 Teachers and academia: The tendency for monolingual communication
A good teacher has not only a wide range of methodological tools at her command, but also becomes very proficient in the subject she teaches. For language teachers, foreign languages are difficult to learn and easy to forget without sufficient practice. Even native speaker language skills may become rusty if not used for a long time. However, it seems that relatively few Japanese teachers welcome the opportunity to practice their language skills, and some avoid it at all costs. It is usually taken for granted, at least in Europe, that a conference about English language or teaching, and probably even about English literature would be held in English, no matter in which country the conference was organized. So, a French scholar of English literature could attend a conference in Bulgaria, Hungary, or Austria and be able to understand it. This is not so in Japan. Surprisingly, in most cases, conference presentations, newsletters, and homepages of scholarly organisations researching foreign languages are in Japanese. One drawback of this is that Japanese teachers who can only rarely attend conferences abroad have few chances to further develop their language skills in an academic environment. Another is that most research conducted in Japan is only accessible to people who have a good command of spoken and written Japanese. This means that many foreign teachers living and working in Japan are effectively barred from taking part or cooperating in research in Japanese. Further, the wider international community has little opportunity to get acquainted with what is going on in the academic world in Japan.

4. Benefits of corpus approach and tools for EFL in Japan

In Section 2.3, corpus linguistics in CALL was described as occurring in four stages (see Fig. 1). Obviously, if teachers want students to use corpus tools, they have to start off by using corpora and corpus tools themselves, and this is probably how most do start – answering their own questions about language. Especially in the case of non-native speaker teachers, these first two steps are very similar in nature to what they will expect their students to be doing in the fourth step; that is to say, improving language skills. Unlike rote learning, this way of studying is very similar to research. Student-researchers discover facts about the foreign language instead of learning ready-to-consume rules proclaimed by the teacher. They observe language data on the basis of which they draw conclusions. By acquiring the skill of being able to spot clues in a context, Japanese students will learn to make intelligent guesses and will thus overcome the paralysing effect of being unable to fill the gaps left by partial understanding.

4.1 Observation

As rote learning has done more to make Japanese students dislike English than help them become fluent in it, learning by discovery will not only develop skills transferable to fields other than language learning but also raise motivation, crucial for learning in general. Key word in context (KWIC) concordancers provide an excellent visual arrangement for observation. As the words can be sorted in many different ways just by pressing a button, new patterns may become easier to notice this way. The use of the computer is much more effective than a printed version of the concordance lines, where the choice has been limited to one, and that has also been chosen by the teacher.

Even low-level students can carry out some observation tasks. A very simple task would be to check in a corpus which expression is more usual in English: "black and white" or "white and black." One might say it is obvious, but it is not so for the language learner. If you were to say *black and white TV* in Japanese or in Hungarian, would you say it the "obvious way"? If so, you would be wrong in the first case and right in the second. *Black and white* can be found in a dictionary; however, no information can be found about the preferred word order in the case of *husband and wife* or *wife and husband*. A corpus search gives a clear indication: the former occurred 144 times in the COBUILDDirect corpus, the latter only 4 times.

With more advanced students, tasks can become more difficult and complex. For example, they can be asked to find sentences with a particular verb in the foreign language corpus, and give possible translation equivalents in their mother tongue. This activity can help students understand that one word in a language can have many different translation equivalents in another language. That would help prevent sentences like "*The computer is not moving," which can often be heard from Japanese students. In this case, the students simply translated the Japanese sentence with the first entry for *ugoku* – the Japanese verb used in this sentence – in their mental lexicon. This kind of observation task can help error correction and raise consciousness at any level. Its positive effect can be increased if the teacher chooses the corpus used for the task carefully. Someya (2000) reports that the use of an online concordancer with a business letter corpus improved overall quality of written messages of Japanese learners of English. Someya's research focused on article and preposition use.

4.2 Drawing conclusions

Observation is usually followed by drawing conclusions, so they go hand in hand. Children learning their mother tongue observe language around them, and create their own lexico-grammar by drawing conclusions. Sometimes these conclusions may lead to mistakes but as they encounter more data, they constantly "revise" their rule system. Instead of memorising rules from grammar books or words from a bilingual word list or dictionary, language learners should try to follow the same procedure. Drawing correct conclusions requires careful observation and categorization. If we look at our previous example of *black and white* and *husband and wife*, a learner may conclude that in some cases the word order is fixed for seemingly independent words. Students becoming aware of such possibilities will approach other expressions with more caution and not jump to the "obvious" solution.

4.3 Intelligent guessing

Intelligent guessing means finding and using contextual clues to help identify missing information. "Missing information" does not necessarily mean that it is actually not there. For example, when reading or listening, although a word may be there, or clearly heard, if the student has no access to its meaning, it is effectively missing. Any communication can be "noisy." Oral communication can be hindered or disturbed by different noises: a passing vehicle, somebody shouting, etc. Written communication can be difficult to read because of print quality or handwriting. In a foreign language, lexical items and syntactical patterns can function the same way as noise described above. As not all elements have the same importance, comprehension also depends on which elements are lost or partially understood. Some students seem not to realise this, either when reading or when listening. This "cluelessness" seems more apparent in listening tasks but reading skills are also negatively affected. Clues can be purely linguistic or can come from the students' knowledge of the world. In most cases both kinds of knowledge are required to solve the problem.

Multiple-choice questions do not foster intelligent guessing. In real life situations, when part of the utterance is not heard, that information is missing, and there might be many words or expressions that could fill the gap. However, the student is not presented with one right and three wrong words or expressions to choose from. The listener has to think of as many possibilities as he or she can in a limited time, and choose the one or ones that make sense in that particular situation and fit linguistically.

Using a corpus and corpus tools, guessing activities can range from simple word lists (Appendix A) to very challenging activities related to lexico-grammatical problems on hard copies or on a computer screen. Advantages of using a computer are that students can have instant feedback and the activity can be linked to a corpus or a set of concordance lines pre-selected by the teacher, where students can look up these expressions in context.

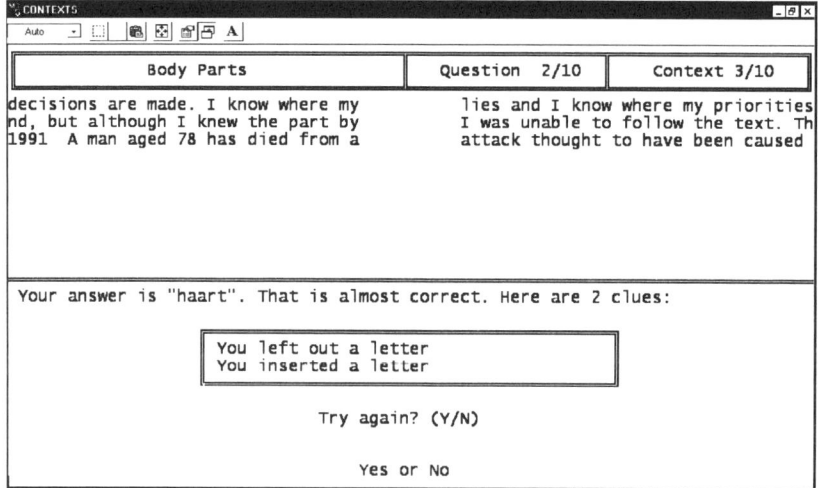

Figure 3. Screenshot from *Contexts.*

The computer program *Contexts* (Johns, 1994) is an excellent tool for practising not only observation but also guessing (Fig. 3). Unlike many CALL programs, the answer is not automatically rejected if there is a spelling mistake, resulting in a higher sense of achievement. However, the program registers all mistakes and reports them to the student on finishing the activity. Another advantage of the program is that it can be tailored by teachers to satisfy their students' special needs. In the case of low-motivated students, this game-like challenge is likely to raise interest.

4. Who could benefit?

The most urgent need for introducing the *learning languages by discovery* approach, using corpora and corpus tools, seems to be in teacher training. Would-be teachers could personally benefit in many ways: They would probably become even more motivated and successful language learners; they could learn some basic skills also needed for research activities; and they would have first-hand experience of the methods they could use later on in their own classes. The growing number of teachers with such experience could then challenge the prevailing belief in rote learning.

As translators and interpreters need to master foreign languages at a high level, they could obviously benefit the same way as trainee-teachers. In addition, they have to learn how to handle the various software because translation has virtually become impossible without the use of corpora, corpus tools, and other computer programs. The need for such instruction is clearly indicated by such conferences as Corpus Use and Learning to Translate organized in 1997 and 2000 by the School for Interpreters and Translators of Bologna University.

In Section 4, several examples were given about how concordancing could be used for very simple tasks. Although it is possible to create material for low-level students, more advanced students could benefit more from such an approach. If students are not able to use a monolingual EFL dictionary, which can be regarded as a kind of edited corpus using a limited number of words for explanation, the challenge of using a corpus of authentic data may be too demanding.

Language learners in any country could benefit from a corpus approach to improve language skills. Japanese students might benefit even more so because it means changing learning habits and becoming less reliant on rote learning. The skills described in Section 4 all require active involvement in the learning process as no rules are presented to the learner.

5. Obstacles and overcoming them

As mentioned previously, low-level students could only benefit if the teacher is willing to sacrifice considerable time either to editing an existing corpus or building one to suit students' special needs. Except for those majoring in languages, considering the generally low level of knowledge of English by Japanese students, it seems premature to expect the wide use of corpora either in high schools or at universities.

Access to computer labs may also be limited, especially below university level. In that case, languages are usually not top priority – science classes are more likely to be scheduled for those labs. In addition, students' computing skills are not as high as one might expect. The large number of students in language classes makes it even more difficult for the teacher to help with computing matters on an individual basis.

Learning habits may be the most difficult to change. Students used to a certain teaching and learning style may expect the same and regard a new style with suspicion or even hostility. In Japan, students usually expect teachers to "teach" them and not to get them to learn something for themselves, an expectation not restricted to learning languages.

These obstacles can be overcome easily. Firstly, teacher cooperation, especially resource sharing (not only corpora but also related activities) can reduce overall class preparation time. As students get used to corpus activities, and enjoy meeting intellectual games and challenges, their learning style can slowly be changed. The popularity of the Internet brought about positive signs of progress in students' computing skills. With the spread of relatively cheap personal computers and Internet services, students are also able to use corpus tools when studying autonomously once introduced.

6. Examples of actual use

6.1 Consultation and teaching materials

Without conducting a nationwide survey, it would be difficult to tell accurately what percentage of English teachers use corpora and corpus tools to improve their own language skills or prepare teaching materials. There are some indications of it becoming a matter of course, as this kind of activity is no longer reported. With the availability of corpora on CD-ROM and online access to corpora, most teachers can use them relatively easily. Two books published in 1998 (Saito, Nakamura, & Akano, 1998; Takaie & Suga, 1998) have played an important part in making the theory and practice of corpus linguistics more accessible for Japanese language teachers specialising in languages other than English. *Eigokyouiku: The English Teachers' Magazine* appears monthly and reaches a wide audience with its corpus related articles and explanations in the "Question Box" section. As most of this publication is in Japanese, it may also be consulted by teachers of other languages, making it possible to apply corpus methods and tools for teaching other languages as well.

The increasing number of research articles together with the number of Japanese authors, and also by people living and working in Japan is an indication of the growing interest in the use of corpora. As many authors are teachers themselves, we can assume that at least some use their findings and tools in creating teaching materials for classes. For example, some websites, such as Peter Ruthven-Stuart's at <www.nsknet.or.jp/~peterr-s/>, offer advice on using concordances in the classroom.

Returning to Figure 1, the four stages do not only represent the process of involvement in corpus-based CALL but also the number of teachers involved in each step. As no concrete figures are available, it just indicates the tendency that the number of teachers consulting corpora or corpus-based resources is much higher than that for teachers making materials. Finally, at present, there may be only a handful of teachers whose classes really use corpus-based CALL.

6.2 Student use of corpora and corpus tools

Although students may not be aware that they are using corpus-based products, they do, like teachers, consult dictionaries on CD-ROM, which may also include some kind of corpus. An example is *Collins Cobuild on CD-ROM*, which contains five million words of authentic texts from the Bank of English corpus.

Concordancing is also widely used by students. A good example of its interactive use can be seen on Ruthven-Stuart's website in quiz form. Students are presented with 10 words and sets of concordance lines with the search word missing. Students need not type but simply drag words to the right concordance lines. The words turn red or green depending whether the answer is correct or not. Ruthven-Stuart also provides information about availability of programs needed to make these quizzes.

A growing number of teachers, like Someya, encourage students to consult a carefully designed corpus in order to improve productive skills. Teacher aims may vary from helping students acquire general vocabulary, technical terms, or styles of certain genres to teaching syntax or linguistics. English for special purposes (ESP) is an area where even a relatively small corpus can be well exploited. Besides online concordancing, several universities use the program called TXTANA in their classes, with one teacher reporting using it in his course on English syntax (T. Nishimura, personal communication, September 17, 2001).

Parallel concordancing (Fig. 4), containing texts in two or more languages, is also becoming popular among Japanese teachers. As translation has always been a focal point in language education in Japan, it fits nicely into this traditional way of looking at languages. Several Japanese universities are engaged in building parallel corpora, used for teaching both Japanese and English.

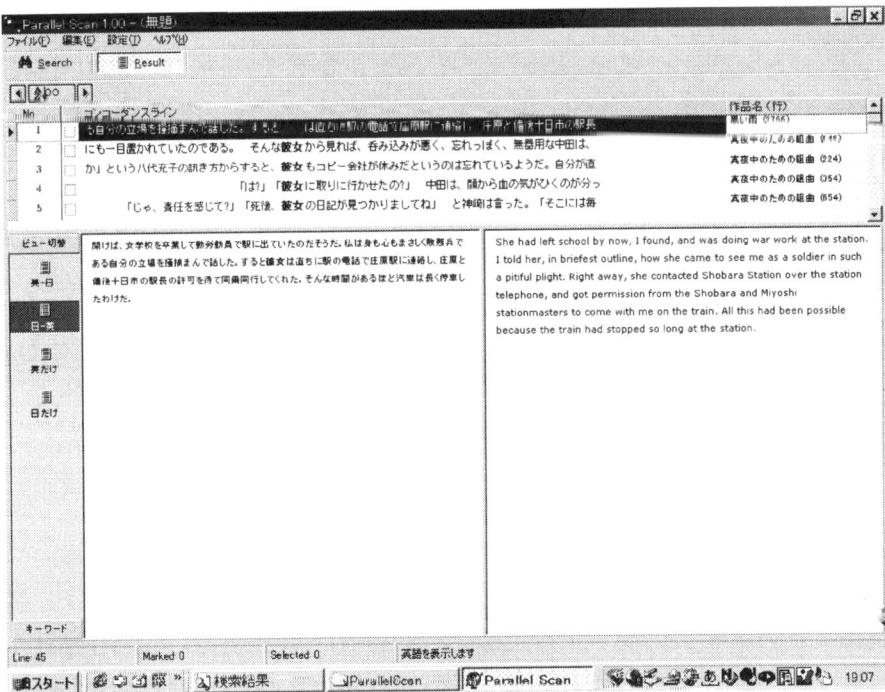

Figure 4. Parallel concordancing; screenshot from *Parallel Scan.*

Among many other possibilities, consciousness raising and comparative studies of English and Japanese seem obvious applications of such a corpus.

6.3 Japanese corpus studies and CALL

Our overview would not be complete without a brief description of corpus studies with respect to the Japanese language. Many research activities carried out by The National Language Research Institute required the collection of linguistic data even decades ago. They installed their first computer system in 1961 in order to investigate "the words and Chinese characters in modern newspapers and the use of such equipments in the linguistic analysis of Japanese" (*An Introduction to the National Language Research Institute*, 1999, p. 102). According to their report, the first project in which the computer was used for data processing resulted in the publication of *Studies on the Vocabulary of Modern Newspapers, Vol 1-4,* the first volume of which appeared in 1970.

In the 1980s, research on the vocabulary of senior and junior high school text-books was carried out "for the purpose of describing and analysing the system of vocabulary, writing forms, and expression in written texts" (*An Introduction to the NLRI*, 1999, p. 113). The concordance with the context "allows the reader to examine the context in which each word was used, and it is useful not only for the specific analysis of words which are used in high school textbooks but also for general lexical and grammatical research on Japanese vocabulary" (p. 114).

In the 1990s, television broadcasts were used for further linguistic research, including spoken data. Concordances of spoken language were created, and more linguistic data gathered under different headings, such as discourse words, spoken language sentence patterns, etc. All original data is publicly available, mostly on microfiche.

It has been an international phenomenon that corpus tools were first used exclusively for research, and became available for language teachers later. However, the use of corpora in learning Japanese seems to lag behind. It may be due to the fact that concordancers able to handle Japanese have not been easy for individuals to obtain. (It took the author almost six years.)

As reports about concordancing programs in the field of teaching Japanese as a foreign language have not been available, the author must rely on personal experience as a learner of Japanese. In addition to that written about the use of concordancing programs in English classes, the learning of reading Japanese should be added. As one Chinese character may have many different readings according to the word or expression it forms part of, or the sense in which it is used, student exposure to a great number of examples provides an occasion for practicing reading as well. In addition, as monolingual dictionaries for Japanese as a foreign language are not available, a corpus search can be the best way to present students with a considerable amount of authentic examples of usage in the shortest possible time.

Morphologic analysers (see Section 2.4) have the potential to become just as popular in teaching agglutinating languages, like Japanese, as concordancing programs are now in teaching English. Combining both could give an instantaneous analysis of each concordance line just as pressing a button can now bring up longer contexts. Figures 5 and 6 are screenshots of morphological analysis of the same sentence. Even without understanding the labels, the difference in organisation of information is obvious. Figure 6 has more and simultaneously less information on screen. When using the interface shown in Figure 6, the viewer has to move the pointer over the word or word combination to see how it is read, while in Figure 5, all readings are shown at the same time. As a growing number of students learn Japanese as a foreign language worldwide, and more and more students come to Japan to study (MEXT, 2001), such programs are certainly going to become very popular due to their effectiveness.

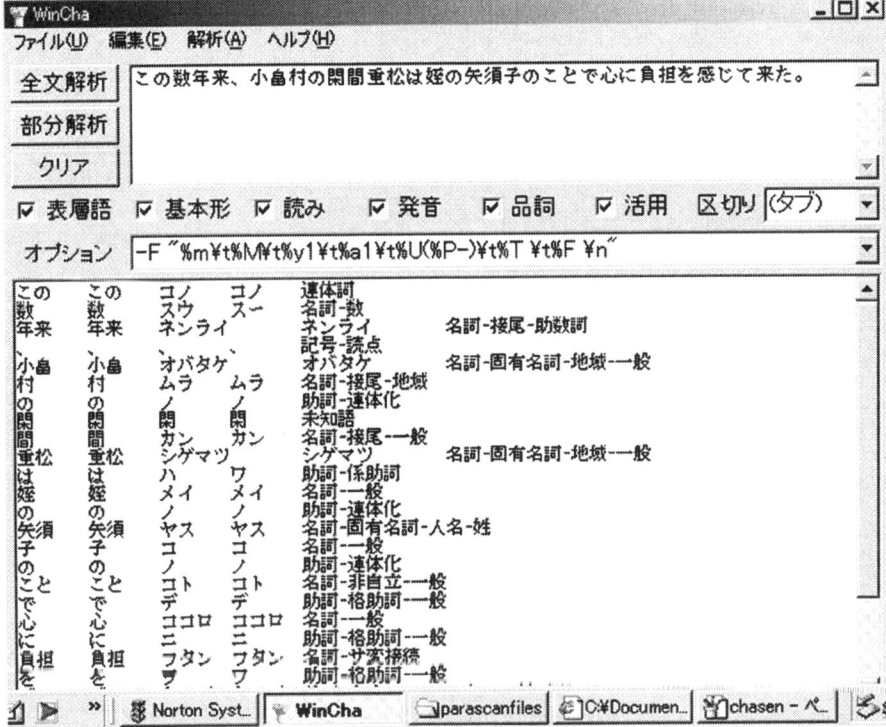

Figure 5. Morphological analysis by Chasen for Windows.

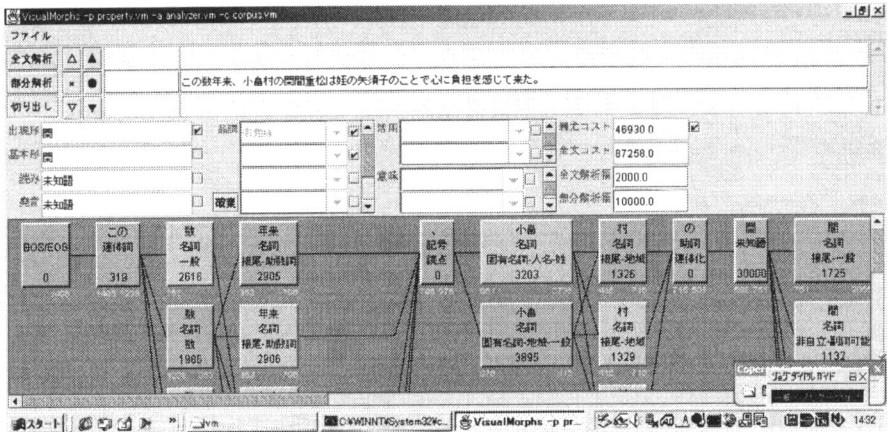

Figure 6. Morphological analysis with visual output.

7. Conclusion

Corpus methods and tools have primarily been used for linguistic research. However, the need for such conferences as the International Conference on Teaching and Language Corpora (TALC), the Practical Applications in Language Corpora (PALC) or the North American Symposium on Corpus Linguistics and Language Teaching show the growing interest in using the results, methods, and tools in language teaching. In Japan, such a conference is yet to be organised, although it has been a focus of JAECS conferences. The two most important results of such a conference are that it could bring together people already using corpus-based CALL and could also make their ideas, experiences, and results known to a wider audience.

In conclusion, it seems that corpus-based CALL has not been widely accepted and used despite the long tradition of corpus studies in Japan. Although the term is frequently used, its meaning is not always clear for teachers. As a result, relatively few people have realised its possibilities other than for research, namely in language teaching. Language education in Japan can only benefit from corpus-based CALL. By introducing corpus-based teaching into teacher training programs, the next generation of teachers could improve their own language skills while bringing different learning and teaching styles to their schools, leaving rote behind. Its true potential is succinctly summarized by Glyn Holmes (1999, p. 239), who called it "one of the bright young stars of CALL."

References

An introduction to the National Language Research Institute: A sketch of its achievements (4th ed.). (1999). Tokyo: National Language Research Institute. (Also available from www.kokken.go.jp/public/eibun50/inl98v1a.pdf)

Barnbrook, G. (1996). *Language and computers: A practical introduction to the computer analysis of language.* Edinburgh: Edinburgh University Press.

Biber, D., Conrad, S., & Reppen, R. (1998). *Corpus linguistics: Investigating language structure and use* (1st ed.). Cambridge: CUP.

Clark, G. (2000, January 30). Why Taro can't speak English. *The Japan Times,* p. 19.

Deschamps, J.-L. (1992). Towards classroom concordancing. In H. Béjoin & P. J. L. Arnaud (Eds.), *Vocabulary and applied linguistics* (pp. 167-181). London: Macmillan Academic and Professional.

Holmes, G. (1999). Corpus CALL: Corpora in language and literature. In K. Cameron (Ed.), *CALL: Media, design, & applications* (pp. 239-270). Lisse: Swets & Zeitlinger.

Housser, R. (1999). Foundations of computational linguistics: Man-machine communication in natural language. Berlin: Springer.

Japan Information Network. (n.d.). *Cram school enrollment.* Retrieved September 20, 2001, from www.jinjapan.org/stat/stats/16EDUA1.html

Johns, T. (1991). Should you be persuaded: Two samples of data-driven learning. *ELR Journal* (4), 1-16.

Johns, T. (1994). Contexts (Version 0.74) [Computer software].

Kennedy, G. (1998). *An introduction to corpus linguistics* (1st ed.). London: Longman.

Konishi, T., Minamide, K., Harakawa, H., & Hachimura, S. (Eds.). (1994). *Genius English-Japanese dictionary*. Tokyo: Taishukan.

Levy, M. (1997). Computer-assisted language learning: Context and conceptualization. Oxford: Clarendon Press.

McEnery, T., & Wilson, A. (1996). *Corpus linguistics*. Edinburgh: Edinburgh University Press.

MEXT Ministry of Education, Culture, Sports, Science and Technology (May, 2001). *Outline of the student exchange system in Japan*. Retrieved September 20, 2001, from www.mext.go.jp/english/kokusai/index.htm

Roosmaa, T., & Prószéky, G. (1998). GLOSSER – Using language technology tools for reading texts in a foreign language. In S. Jager, J. Nerbonne & A. v. Essen (Eds.), *Language teaching and language technology* (pp. 101-107). Lisse: Swets & Zeitlinger.

Ruthven-Stuart, P. (2000). Concordances: Production and use for teachers and learners. In P. N. D. Lewis (Ed.), *Calling Asia: The proceedings of the 4th annual JALT CALL SIG conference Kyoto, Japan* (pp. 177-182). Nagoya: Chubu Nihon Kyouiku Bunkakai.

Saito, T., Nakamura, J., & Akano, I. (1998). *Eigo kopas gengogaku: Kiso to jissen.* [English corpus linguistics: Foundation and practice]. Tokyo: Kenkyusha.

Sinclair, J. (Ed.). (1987). *Collins COBUILD English Language Dictionary.* London: Collins.

Sinclair, J. M. (2001). Preface. In M. Ghadessy, A. Henry, & R. L. Roseberry (Eds.), *Small corpus studies and ELT: Theory and practice* (vii-xv). Amsterdam; Philadelphia, PA: John Benjamins.

Someya, Y. (2000). *Online business letter corpus KWIC concordancer and an experiment in data-driven learning/writing.* [Web document]. Retrieved September 17, 2001, from ftp.kamakuranet.ne.jp/~someya/DDW_Report.html

Studies on the vocabulary of modern newspapers (vols. 1-4). (1970-1973). Tokyo: Shuei Shuppan.

Szirmai, M. (2001). Corpus tools in language teaching and learning. In J. White (Ed.), *FLEAT IV (The 4th Conference on Foreign Language Education and Technology)* (pp. 146-151). [CD-ROM]. Kobe, Japan: The Japan Association for Language Education and Technology.

Takaie, H., & Suga, H. (1998). *Jissen kopas gengogaku: Eigokyoshino intanetto katsuyo.* [The practice of corpus linguistics: The English teacher's use of the Internet]. Tokyo: Kirihara Yuni.

Takebayashi, S., Kojima, Y., & Higashi, N. (Eds.). (1996). *Lighthouse English-Japanese dictionary* (3rd ed.). Tokyo: Kenkusha.

Takefuta, Y. (1981). *Kompyutano mita gendai eigo: bocabyurarino kagaku.* [Modern English on computer: The study of vocabulary]. Tokyo: Eduka.

TOEFL test and score data summary. (n.d.). Retrieved September 20, 2001, from www.toefl.org/pubs/resdloadlib.html#summaries

Tribble, C., & Jones, G. (1990). *Concordances in the classroom.* London: Longman.

Websites and software

Note: Some websites are bilingual, some only Japanese or English. However, even the Japanese websites may have links to useful information in English. To view Japanese websites, change the "Character Set" or "Encoding" of your browser, which can be found in the "View" menu, to Japanese (Auto-Detect).

AOZORA Online electronic texts in Japanese: www.aozora.gr.jp/
ARIADNE Japanese Literature: ariadne.ne.jp/literature-j.html
Business letter corpus and online KWIC concordancer:
isweb9.infoseek.co.jp/school/ysomeya/
Chasen: chasen.aist-nara.ac.jp/
Gotoo Hitoshi's links to linguistics and language-related websites in Japan:
www.sal.tohoku.ac.jp/~gothit/kanren-en.html
Japan Association for English Corpus Studies: muse.doshisha.ac.jp/JAECS/
KWICK concordance for Windows: www.chs.nihon-u.ac.jp/eng_dpt/tukamoto/
kwic.html
Peter Ruthven-Stuart's homepage: www.nsknet.or.jp/~peterr-s/
Software tools for NLP: www-a2k.is.tokushima-u.ac.jp/member/kita/NLP/
nlp_tools.html
TXTANA: www.biwa.ne.jp/~aka-san/index.html
Yukio Tono's homepage: www.lancs.ac.uk/postgrad/tono/
(*mirror site*: www.lb.u-tokai.ac.jp/tono/default.htm)
VisualMorphs: chasen.aist-nara.ac.jp/stable/doc/macd/NL20000601.files/frame.htm

Appendix A

Guessing game
Activity 1: Look at the 4 lists. Each list contains the words that most often appear together with the following 4 words: **under, take, turn, hand.** Can you find which of these 4 words goes with each list?

other	7872
right	2752
left	2150
held	1335
second	1276
put	1186
first	1119
back	996
made	755
took	676
side	664
down	579
go	579
upper	529
man	525
painted	487
holding	471
off	457
got	439
free	426

place	7398
off	4990
time	4067
part	3962
action	3219
care	3154
more	3095
some	2851
away	2425
advantage	2359
long	2350
going	2238
any	2178
people	2159
back	2033
look	1949
account	1902
just	1820
years	1724
now	1674

Activity 2: In each list, underline the words that helped you make up your mind. Compare your underlined words with somebody else's. Are they the same?

around	1395
century	1366
off	1337
back	1227
down	1068
right	903
now	829
left	775
round	622
away	613
just	562
every	482
might	453
each	402
page	385
took	379
going	369
made	333
things	326
attention	323

pressure	5395
control	4001
way	3729
new	2918
circumstances	1935
come	1934
now	1924
law	1860
government	1854
just	1790
conditions	1749
fire	1595
came	1464
par	1392
attack	1380
children	1309
put	1298
threat	1287
system	1212
name	1193

(The lists were taken from *Cobuild English Collocations on CD-ROM*.)

8

Giving Learners Access to Vocabulary Enrichment

Bradley Saunders
Zayed University, Dubai, UAE

1. Introduction

The teaching of vocabulary has only recently emerged from a long period of neglect in the ELT literature, and indeed faces new opportunities as recent insights from both computer-based corpora research and psycholinguistic studies of the lexicon challenge traditional approaches to teaching lexis (McCarthy, 1990). Recent thinking in vocabulary teaching is tending towards a dynamic, process-centred approach. Vocabulary learning is increasingly seen as a skill and responsibility lies with the learner to develop his or her own lexicon.

This is not to say however, that teachers can avoid responsibility for helping their students to consolidate and extend their vocabulary. As Nation (1990, p. 1) points out in the introduction to one of the most widely-read texts on vocabulary teaching, "it is not difficult to find language teachers who think that vocabulary can be left to take care of itself. " There are several reasons why. Firstly, the recent emphasis on communicative methodology with its implied acceptance of risk-taking and guesswork does not fit well with the "intimidating and authoritarian" (Wright, 1998, p. 10) image of the dictionary. Secondly, as Cobb and Horst (2001) point out, learners bring with them widely different vocabulary needs – a word unknown to one student may well be very familiar to his or her neighbour. Consequently, it is difficult for a teacher to select vocabulary items to cover with students (Gairns & Redman, 1986, p. 55). A third factor contributing to the sidelining of vocabulary teaching is the view that dictionaries and other vocabulary references are best used privately, in individual study.

This paper describes the use of a database program to empower students enrolled on a technical writing course in the College of Information Systems (CIS) to systematically record both technical and general vocabulary items. These students had just completed the first two years of their university education and embarked upon a two-year baccalaureate program in the college. They faced two daunting tasks: absorbing a large number of new technical vocabulary items in such fields as computer graphics, database structures, Internet technology and programming concepts, and reactivating a large number of academic and general vocabulary terms learned in the previous two years of their university careers. A total of 108 students used the program for a period of 12 weeks.

The students in question were attending a women's university in the United Arab Emirates. I will therefore highlight some similarities between their needs and those of Japanese learners of English before examining the issues of how vocabulary is learned, considering the role of CALL in promoting learner autonomy, and discussing the application itself.

2. Learner background

Space does not permit a full discussion and comparison of learning styles and backgrounds of Emirati and Japanese learners of English. However, let us draw some general comparisons to appreciate the relevance of this study to the vocabulary learning needs of Japanese students of English.

Looking first at the learners' linguistic background, there is an obvious parallel in that neither Arabic nor Japanese uses Roman script. This implies problems with "the recognition of the alphabetic system, the correlation of the graphic symbols, as well as intellectual comprehension" (Al-Mutawa & Kailani, 1989, p. 114). Of particular importance to the teaching of vocabulary, Al-Mutawa and Kailani also point out that the teacher of Arabic learners "will not have the advantage of cognates which might facilitate his task of teaching new lexical items" (p. 49). Turning to their educational background, in common with their Japanese counterparts, Emirati students come from:

> An educational system where memorisation is stressed, explanations are lock-step, and mastery is expected through the successful completion of set tasks. (Towndrow, 1996, p. 35)

The directive, prescriptive educational background of Emirati students in this study is similar to that of their Japanese counterparts. Even after two years of study at the tertiary level, most students express a clear preference for teacher-directed activities rather than the self-directed mode that autonomy implies. This was not significantly altered when ICT became a major component of their teaching and learning situation (Radecki, forthcoming).

There are, therefore, clear parallels between the two groups of learners which make the results of this study relevant to a publication of this nature. However, to maximise the benefit of the program to the Japanese learners with whom it will be trialled in the near future, an ongoing dialogue has been established with teachers in

Japan. As Johnson and Brine point out: "all successful developments and implementations of educational computer applications are influenced and shaped by the local context in which they are designed and used and include the entire milieu of cultural and social organisation" (1999, pp. 251-252).

3. Vocabulary notebooks

Students are often advised to take notes on vocabulary items they encounter for the first time in order to facilitate retention. In a vocabulary guide written for students, Lacey et al. dedicate one of eight chapters to techniques of recording vocabulary and advise students to "write words on record cards which can be stored in a box in alphabetical order" (1990, p. 41). Leeke and Shaw describe how the Japanese learners in their study had all at some time used "the Japanese pre-printed vocabulary book. These are notebooks with pages divided into blank columns, each with a heading in English, published in Japan" (2000, p. 278).

Several teachers' guides place importance on the use of notebooks and record keeping (Allen, 1983, p. 50; Gairns & Redman, 1986, p. 95-100; Schmitt & Schmitt, 1995). Woolard states "It is now clear that we need to give vocabulary notebooks a far greater priority in language teaching, and raise our students' awareness of the dynamic role they have to play in the process of learning a language" (2000, p. 44). However, with a few notable exceptions (Ahmed, 1989; Gu, 1994; McCarthy, 1990; Leeke & Shaw, 2000) there has been little research into vocabulary note-taking and its effects on vocabulary learning and retention.

Proponents of the use of vocabulary notebooks or record card systems cite:
- their convenience: "their handy size makes them convenient to carry around and easy to study in odd minutes of free time" (Schmitt & Schmitt, 1995, p. 137).
- their sortability. "one of the advantages of a card index is that cards (words) can be pulled out to make a set . . . . Then you can put the cards back in alphabetical order in their box" (Lacey et al., 1990, p. 45).
- their updatability: "Items in the notebooks are not just listed and left. They are revisited and extended in the light of the learners' increased exposure to the language" (Woolard, 2000, p. 43).

The students in the author's institution did not need to purchase notebooks or index cards to benefit from these advantages. Each one was already in possession of an even more convenient, sortable, and updateable source of vocabulary enrichment, namely their university-issued laptop computer. They only needed to share a common interface to exploit the power of the technology at their disposal.

4. CALL and vocabulary learning

Such use is in line with the increasing trend away from the idea of computer as *tutor* to that of computer as *tool* (Levy, 1997; Susser, 1998). The tutor role is one in which the computer transmits knowledge and skills, whereas in the role of tool, it empowers learners to carry out learning tasks. "The profession now realizes that a computer is a *medium for learning and not a method for L2 instruction*" (Adair-Hauck et al., 1999, p. 272, italics in original).

Clarke (1992) reviews the role of CALL vocabulary programs and bemoans their lack of contextualisation, overemphasis on form – especially spelling – over meaning, and focus on *testing* rather than *teaching*. He argues for the use of computers in an "informative" role in which "learners could be involved in practical tasks of a discourse nature while developing the support systems of the computer, by adding lexical information as it arises in their study of successive texts" (1992, pp. 140-141).

This approach is very similar to the way the software works. However, it is important to remember that the simple act of storing a word – in whatever medium – does not guarantee its retrieval. Factors involved in short and long-term memorisation will determine how well and for how long learners retain words (see Schmitt & Schmitt, 1995, for a review of the important factors). Strengths of the software in this respect are that it allows learners to easily organise vocabulary items into meaningful groups, increases the number of exposures learners receive to vocabulary items, and puts the onus on learners to choose which items to store: " Learning words in another language cannot be easily divorced from motivational factors such as how important or useful lexical items are perceived to be by learners themselves" (Carter & McCarthy, 1988, p. 16).

5. CALL and autonomy

For such a system to work, it is necessary for learners to be comfortable taking on an autonomous role. Autonomy is generally defined as the ability to take charge of one's own learning (Holec, 1981; Little, 1995; Wenden, 1998). Although this necessarily implies a key role for independence and individual responsibility, several researchers stress the importance, in autonomous learning, not only of *independence* but also of *interdependence* (Boud, 1981; Little, 1990; Blin, 1999). This is particularly relevant to the Japanese context. Brajcich (2000) cautions that "it is not always easy to develop learner autonomy in cultures such as Japan's, where forming social groups and *amaeru*, or interdependence, are the accepted norm" while Aoki and Smith (1999) argue that Japanese learners respond well to a group-based autonomous learning approach.

The mere provision of a laptop computer for each student and the concomitant opportunity for self-study does not guarantee autonomy. The success of autonomous learning using CALL, whether on an individual or a group basis, depends on the role of the computer and CALL software being used, attitudes of learners, and the ability of the teacher to promote autonomy in the students.

The computer has to grant decision-making authority to learners. This will not occur if the software used fulfils the role of "tutor." Levy contrasts the use of drill and practice programs, tutorials, simulations, and so on in the role of tutor with conventional application programs such as word processing packages, databases, and telecommunication packages in the role of tool (1997, pp. 180-187). Jones provides further insight into the changing role of CALL programs: "The versatile modern generation of CALL programs, exploiting a gamut of applications from word processing to virtual learning environments, tends to show awareness of the need to give the student significant responsibility of the management tasks in his or her learning" (Jones, 2001, p. 1).

Learners' beliefs have an influence on their ability to develop autonomy. Toyoda (2001) postulates that the more positive learners' attitudes towards technology are, the more they will be able to make use of the technological tools at their disposal to achieve the level of autonomy they require. Jones (2001, p. 2), reminding us that "there are individuals who do not take to the computer as a tool for learning a language," indeed hypothesizes that, by recognising they do not learn effectively through this medium, they are, ironically, proving their capacity for autonomy.

The use of technology to promote student autonomy moves the focus away from the teacher, who must be comfortable devolving the central role and encouraging students to function unaided. McGarrell (1998, p. 138) analyses the changing roles of teachers and learners in the technological classroom and points out that "none of these roles render the instructor obsolete, but they tend to alter existing relationships and dynamics in the language learning environment." Clearly, just as learners can choose not to use the technology at their disposal, so can the teacher choose not to encourage them to do so, whether out of fear of the effect this will have on his or her role or because of lack of confidence with the technology.

6. Program description

On launching the program, users are presented with the screen shown in Figure 1.

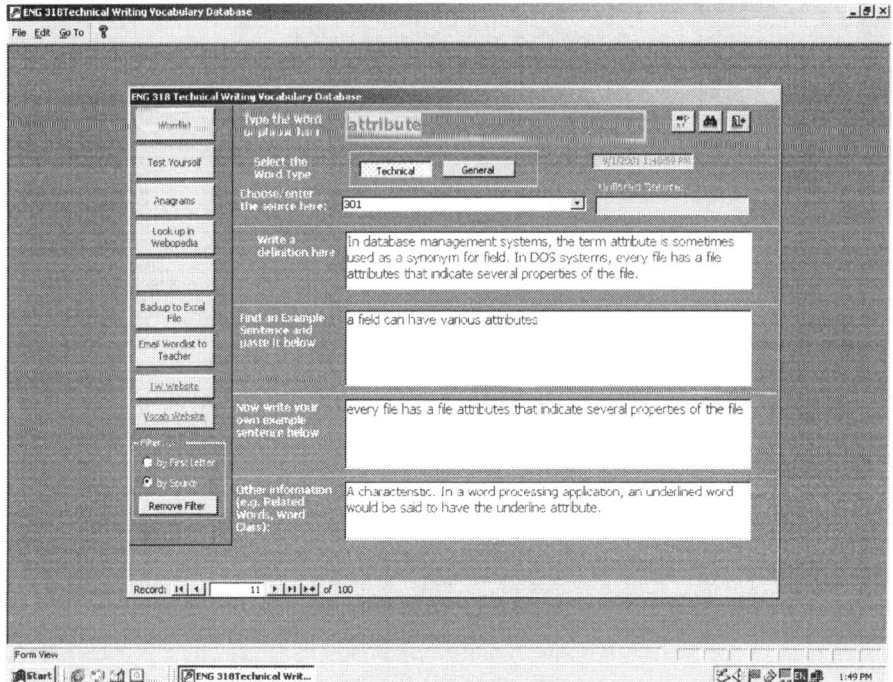

Figure 1. Opening screen.

After entering the word or phrase, they are asked to assign it as technical or general. Depending on the category they choose, the word or phrase they enter is stored as part of a hyperlink to a technical or general English dictionary. This is then called by clicking on the relevant button. Students view and paste in a definition and an example sentence from the online dictionary and are also required to assign the word or phrase to a source. This, in effect, creates a group for the word in which it can later be sorted. In this study, the list of sources consisted of the names of courses held in the CIS. Several other possibilities could have been used, such as thematic areas, chronological groupings, or subject-specific areas. Lewis (1997) sees an advantage in having such groups set up by the teacher and points out that "the range and sophistication of the headings can be tailored to specific classes" (p. 77). However, a degree of flexibility was offered to students by allowing them to enter new sources as required.

In addition to the definition and example sentence obtained from the Web dictionary or another external source, a field was provided for students to write in an example sentence of their own. (This proved to be a challenging aspect of the program; see below for a consideration of issues in grading such sentences.) The final field was intended for notes, including part of speech, typical collocations, and general comments. Students were also able to enter an L1 equivalent in this field if they felt it necessary.

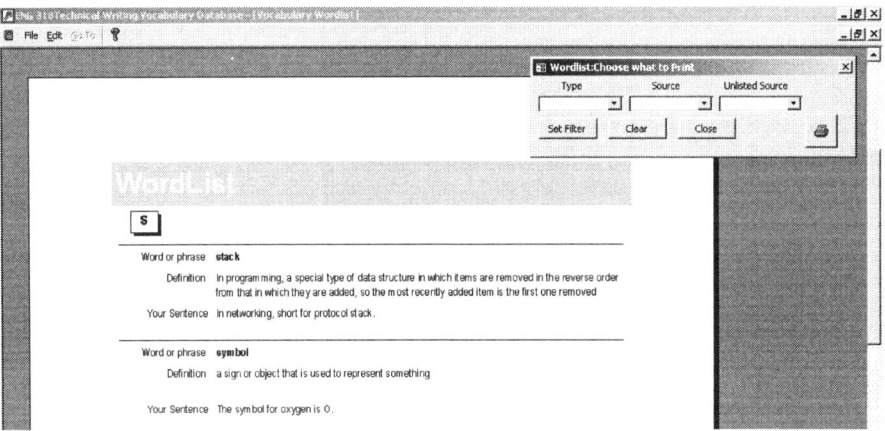

Figure 2. Print filter.

The left-hand side of the screen is given over to several command buttons. The top three (Wordlist, Test Yourself, and Anagrams) share a common interface (Figure 2) which allows users to filter their wordlists according to word type (technical or general) and source (including new sources added by the user) and then print out an alphabetical listing of the words or very simple exercises to help reinforce them. A button was provided for students to email the first of these three documents to their teacher as an attachment in Word rich text format. An equally important function was performed by the backup button, which saved a copy of the wordlist in Excel. This could then be re-imported through a macro listed in the File menu. In three cases, this proved necessary following corruption of the database file.

7. Grading

The program was created with the aim of having students function autonomously. However, it is necessary, as Schmitt and Schmitt (1995, p. 140) point out, for teachers to periodically check over learners' entries since "it does no good to have students efficiently learning errors!" The students in this study, when the vocabulary database was first shown to them and fully explained, indicated very strongly that an activity such as this, on which they were expected to expend considerable time and effort, should be graded. It was therefore decided to dedicate a small percentage of the course grade to student wordlists created in the database. It is a sign of students' lack of autonomy that not only were they motivated by the possibility of a good grade in the vocabulary section of the course but many of them also expected, indeed *demanded*, to know how many and which vocabulary items to enter into their database!

The grading system was based on that proposed by Summers (1988) (see also Cobb & Horst, 2001, p. 208-209). Basically, random samples were taken of each student's sentences. These were then grouped into the following categories (Summers, 1998, p. 121):

- Correct and original (O): The student understood the word meaning and used it in a grammatically correct sentence departing from the definition given.
- Correct (C): The student's illustrative sentence is grammatically correct, but too similar either to the definition given or the comprehension text, or alternatively not informative enough to indicate that they have fully understood the meaning (Table 1).
- Incorrect (I): The student has made an error of syntax, semantics, collocation, etc. (Table 2).

Table 1. Category C errors.

Word	Student Definition
Implement	I had implemented my project
Consumer	We are like a consumer
Cookie	I learned about cookies at the university.

Table 2. Category I errors.

Word	Student Definition
Confident	Although I was confident that I will get low grade in test, I got high grade
Engaged	"Can you come on Monday?" "Sorry, I can't because I'm engaged this week"
Expire	My friend will expire her studying at the university after two years

The database was structured to allow users to enter the same word more than once to accommodate words with more than one meaning. Generally speaking, common words are much more likely to be polysemous than technical words, very many of which are monosemous to reflect precise meanings among professional users. However, a sizeable group of words can be used in both a nontechnical and a technical sense. As Nuttall points out:

> Some of the most dangerous misunderstandings arise when apparently every-day words are used in specialized senses by writers in specialized fields. The mathematician's use of *argument*, the statistician's *random*, the communications expert's *noise*, are all very different from the layman's. (Nuttall, 1982, p. 77, italics in original)

Indeed, several instances were noted of words classified and defined as technical by the students but for which the example sentence exemplified general use or vice versa (see Table 3).

Table 3. Errors of multiple meaning.

Word	Definition and Type	Example Sentence
dim	not bright (General)	Dim can help you in problem solving in IS as variable.
device	a piece of equipment that is used for a particular purpose (Technical)	With both parents out at work, the kids were often left to their own devices.
attached	to like someone or something very much (General)	I attached files to the email.
incorporated	used after the name of companies which have been organized in a particular legal way (Technical)	I incorporated the letter into my diary.

8. Future Enhancements

The program used in this study can at best be described as a beta version. It lacks several refinements of potential benefit to learners. These are discussed below and will hopefully be incorporated into future versions of the program.

8.1 Sound

Research shows that the ability to hear new words pronounced enhances student learning of those words (Stanovich, 1986; Kelly, 1992; Hill, 1998). Greaves (2001, p. 142-146) offers an excellent introduction to the use of text-to-speech software. The slightly unnatural timbre of its synthesized voices is more than compensated for by the flexibility of the text-to-speech engine. While native-speaker recordings would ensure a

better quality model, it would require a huge effort (and a huge hard disk) to anticipate and record any word users might enter.

8.2 Exercises

Informal interviews with students participating in this study seem to confirm the observation made by Cobb and Horst:

> Hulstijn, Hollander, and Greidanus (1996) have shown that the act of looking
> up a word in a dictionary increases the chances that the learner will remember
> it; it seems likely that the act of typing out a definition and example sentence
> also contributes to making the word memorable. (Cobb & Horst, 2001, p. 196)

However, many of the learners in this study were not, especially at first, particularly
highly motivated to use the database. Therefore, it was felt important that they be
offered something more to do other than simply entering the word and its associated
definition and example sentences into the database. As Leeke and Shaw point out:
"often the making of the list is the only processing the word gets and where this is not
enough to fix it in memory the activity begins to seem pointless" (2000, p. 280). The
exercises are intended to increase the number of times learners come across the items
and thus help store them in long term memory (Nation, 1990).

Although, as mentioned above, the exercises are currently paper-based, it is hoped
to develop them into activities that can be completed on screen in the next version of
the program. Whatever media they use, however, the exercises are certainly open to
Clarke's criticisms of failing to address wider contextual issues and of being more
like testing than teaching materials (1992, p. 140). It would appear essential that the
type of exercises generated by the program involve learners in deeper mental processing
(Craik & Lockhart, 1972). Schmitt and Schmitt suggest such activities as "judging
the formality of a word, or grouping the word with other conceptually associated
words" (1995, p. 135). The latter suggestion certainly seems a good candidate for a
future exercise, given the program's filtering capabilities.

Complete knowledge of a word involves several different aspects (Richards, 1976),
and the role of concordancing software in vocabulary acquisition is well documented
(Thurstun & Candlin, 1998; Cobb, 1997). To encourage learners to think beyond
decontextualised item-by-item correspondences, and to further exploit the power of
hyperlinks, it is planned to incorporate a corpus from which students can produce
concordances for the items they enter.

8.3 Dictionary source

The current version of the program is dependent upon third-party web-based dictionaries. Whilst excellent sources of well-written definitions and examples, they are not
under the control of teachers. They cannot be edited directly to add missing terms,
(the learners in this study needed to stay abreast of the rapidly-expanding range of
terms used in such topics as object-oriented programming, Java, and web design),
improve the relevance of a definition, or avoid cultural problems in an example sentence. They can go offline at a minute's notice. Worse, as free resources, they could
disappear tomorrow.

To avoid these problems, it is planned for the faculty in the College of Information Systems, together with language experts from the English Language Centre, to draw up an in-house glossary of technical terms, definitions, and example sentences. If these are generated from existing course materials using concordancing software, a much better "fit" with course objectives can be achieved. For a discussion of the creation of such lists within the Japanese university context and a set of guidelines, see Stapleton (1998).

8.4 Student reaction

Overall, the program was very favourably received by a majority of students. This was due in no small part to its novelty value and also since the participants of the study – students of information systems in general and budding programmers in particular – showed a natural propensity to experiment with the program and attempt to work out its underlying algorithms. Several students expressed their intention to continue using the program after the course had finished. Student comments in informal unstructured questionnaires and interviews, together with instructors' ethnographic observations, have helped successfully eliminate a number of minor bugs and generate a wish list of features for future versions.

Work has started on incorporating many of these features into the program, which should be greatly enhanced as a result. At the same time, it is important not to add features "because we can." Any CALL developer would do well to heed the words of Susser: "Almost every CALL package I have looked at has some element of this: 'features' put in seemingly because the computer can do them rather than for any educational benefit they might have" (Susser, 1998, p. 13).

An additional use to which the program was put, given the fact that the students were on a technical writing course, was the requirement that they author a user manual for future students. The authors of the manual will be asked to train new students on the course to use the application. They, in turn, will be required to author an FAQ and possibly a Windows help file to accompany the manual.

A group of six colleagues is beta testing an adapted version of the software for use with students in the Readiness program of the English Language Centre. Following the presentation of the software at the JALTCALL2001 conference in Ota, Japan in May 2001, the software is also being adapted for use by Japanese learners. It is planned to re-author the application as a shareware program using Real Basic on the Apple Macintosh, which would enable cross-platform running.

At the time of writing, students are about to return from summer vacation. Research is planned to ascertain the number of students who continue to use the program voluntarily and to measure vocabulary gains brought about by its use.

References

Adair-Hauck, B., Willingham-McLain, L., & Youngs, B. E. (1999). Evaluating the integration of technology and second language learning. *Calico Journal, 17*(2), 269-306.

Ahmed, M. O. (1989). Vocabulary learning strategies. In P. Meara (Ed.), *Beyond words* (pp. 3-14). British Association for Applied Linguistics, in association with the

Centre for Information on Language Teaching and Research.

Al-Mutawa, N., & Kailani, T. (1989). *Methods of teaching English to Arab students.* Harlow: Longman.

Allen, V. F. (1983). *Techniques in teaching vocabulary.* Oxford: OUP.

Aoki, N. & Smith, R. (1999). Autonomy in cultural context: The case of Japan. In S. Cotterall & D. Crabbe (Eds.), *Learner autonomy in language learning: Defining the field and effecting change* (pp. 19-28). Frankfurt am Main: Lang.

Blin, F. (1999). CALL and the development of learner autonomy. In R. Debski & M. Levy (Eds.), *WorldCALL: Global perspectives on computer-assisted language learning* (pp. 133-147). Lisse: Swets and Zeitlinger.

Boud, D. (Ed.). (1981). *Developing student autonomy in learning.* London: Kogan Page.

Brajcich, J. (2000). Encouraging learner autonomy in your classes. *TLT Online.* Available from langue.hyper.chubu.ac.jp/jalt/pub/tlt/00/mar/sh_brajcich.html

Carter, R., & McCarthy, M. (Eds.). (1988). *Vocabulary and language teaching.* London: Longman.

Clarke, M. (1992). Vocabulary learning with and without computers: Some thoughts on a way forward. *CALL, 5*(3), 139-146.

Cobb, T. (1997). Is there any measurable learning from hands-on concordancing? *System, 25*, 301-315.

Cobb, T., & Horst, M. (2001). Growing academic vocabulary with a collaborative online database. In B. Morrison, D. Gardner, K. Keobke, & M. Spratt (Eds.), *ELT perspectives on IT & multimedia.* Hong Kong: English Language Centre, The Hong Kong Polytechnic University.

Craik, F., & Lockhart, R. (1972). Levels of processing: A framework for memory research. *Journal of Verbal Learning & Verbal Behaviour, 11*, 671-684.

Gairns, R., & Redman, S. (1986). *Working with words: A guide to teaching and learning vocabulary.* Cambridge: CUP.

Greaves, C. (2001). Resource-assisted learning: A vocabulary on-demand model for WWW online learning. In B. Morrison, D. Gardner, K. Keobke, & M. Spratt (Eds.), *ELT perspectives on IT & multimedia.* Hong Kong: English Language Centre, The Hong Kong Polytechnic University.

Gu, Y. (1994). Vocabulary learning strategies of good and poor Chinese EFL learners. In N. Bird, P. Falvey, A. B. M. Tsu, D. M. Allison, & A. McNeill (Eds.), *Language and learning* (pp. 376-401). Hong Kong: Hong Kong Education Department (ERIC Document Reproduction Service No. ED 370 411).

Hill, M. M. (1998). Words in your ear: English vocabulary learning program. In C. McBeath, C. McLoughlin, & R. Atkinson (Eds.), *Planning for progress, partnership, and profit.* Perth: Australian for Educational Technology, Proceedings EdTech'98.

Holec, H. (1981). *Autonomy and foreign language learning.* Oxford: Pergamon.

Hulstijn, J. H., Hollander, M., & Greidanus, M. (1996). Incidental vocabulary learning by advanced foreign language students: The influence of marginal glosses, dictionary use, and reoccurrence of unknown words. *The Modern Language Journal, 80*, 327-339.

Johnson, E. M., & Brine, J. W. (1999). Design and development of CALL Courses in Japan. *CALICO Journal, 17*(2), 251-268.

Jones, J. (2001). CALL and the teacher's role in promoting learner autonomy. *CALL-EJ Online*. Available from www.lerc.ritsumei.ac.jp/callej/6-1/jones.html

Kelly, P. (1992). Does the ear assist the eye in the long-term retention of lexis? *IRAL, 18*(2), 137-145.

Lacey, C., Mahood, J., Trench, J., & Vanderpump, E. (1990). *Increase your vocabulary*. Oxford: OUP.

Leeke, P., & Shaw, P. (2000). Learners' independent records of vocabulary. *System, 28*(2), 271-289.

Levy, M. (1997). *Computer-assisted language learning: Context and conceptualization*. New York: OUP.

Lewis, M. (1997). *Implementing the Lexical Approach*. Hove: Language Teaching Publications.

Little, D. (1990). Autonomy in language learning. In I. Gathercole (Ed.), *Autonomy in Language Learning* (pp. 7-15). London: CILT.

Little, D. (1995). Learning as dialogue: the dependence of learner autonomy on teacher autonomy. *System, 23*(2), 175-182.

McCarthy, M. J. (1990). *Vocabulary*. Oxford: OUP.

McGarrell, H. M. (1998). The computer in English as a subsequent language writing: Roles & relationships. In P. N. D. Lewis (Ed.), *Teachers, learners, and computers: Exploring relationships in CALL*. Nagoya: Chubu Nihon Kyouiku Bunkakai.

Nation, I. S. P. (1990). *Teaching and learning vocabulary*. Boston, MA: Heinle & Heinle.

Nuttall, C. (1982). *Teaching reading skills in a foreign language*. London: Heinemann.

Radecki, W. (forthcoming). *Student and teacher preferences in the high-tech classroom*.

Richards, J. (1976). The role of vocabulary teaching. *TESOL Quarterly, 10*, 77-89.

Schmitt, N., & Schmitt, D. (1995). Vocabulary notebooks: Theoretical underpinnings and practical suggestions. *ELT Journal, 49*(2), 133-143.

Stanovich, K. (1986). Matthew effects in reading: Some consequences of individual differences in the acquisition of literacy. *Reading Research Quarterly, 21*(4), 360-406.

Stapleton, P. (1998, January 18). Faculty word bank on the Internet. *The Language Teacher Online*. Retrieved August 31, 2001, from langue.hyper.chubu.ac.jp/jalt/pub/tlt/98/jan/stapleton.html

Summers, D. (1988). The role of dictionaries in language learning. In R. Carter & M. McCarthy (Eds.), *Vocabulary and language teaching*. London: Longman.

Susser, B. (1998). CALL teacher education: Components and questions. In P. N. D. Lewis (Ed.), *Teachers, learners, and computers: Exploring relationships in CALL*. Nagoya: Chubu Nihon Kyouiku Bunkakai.

Thurstun, J., & Candlin, C. N. (1998). Concordancing and the teaching of the vocabulary of academic English. *English for Specific Purposes, 17*(3), 267-280.

Towndrow, P. A. (1996). *The role and utility of CALL in ELT: A case study of a first level course at the UAE University*. Unpublished master's dissertation, University of Surrey.

Toyoda, E. (2001). Exercise of learner autonomy in project-oriented CALL. *CALL-EJ Online, 2*(2). Available from www.lerc.ritsumei.ac.jp/callej/5-2/toyoda.html

Wenden, A. L. (1998). *Learner training in foreign/second language learning: a curricular perspective for the 21st century (ED 416 673.).* ERIC Reproduction Services.

Woolard, G. (2000). Collocation – Encouraging learner independence. In M. Lewis (Ed.), *Teaching collocation* (pp. 28-46). Hove: Language Teaching Publications.

Wright, J. (1998). *Dictionaries: Resource books for teachers.* Oxford: OUP.

9

Assessing the Potential of Computerised Bilingual Dictionaries for Enhancing English Vocabulary Learning

John Paul Loucky
Seinan Women's University

1. Introduction

Despite the benefits of rapid access computerised bilingual dictionaries (CBDs), not all Japanese college students use them yet. Few studies of computerised dictionaries have been done in Japan, and no published study comparing bilingual portable devices, translation websites, or software could be found. Many teachers and learners are unaware of this technology's potential as an aid in language teaching and learning. Since Perry's (1997) study covered only monolingual electronic learners' dictionaries (ELDs), this study examined other types of electronic dictionary. It grew out of a dissertation comparing CAI with traditional text-based and audiolingual methods of vocabulary instruction. Then, a replication of Laufer and Hadar's 1997 study compared monolingual with bilingual book dictionary definitions, and finally a comparative study was made of *kanji* (Chinese characters) and English vocabulary training approaches, systems, and strategies (Loucky, in press a).

The short-term goal of this study was to determine the relative effectiveness of various dictionary-based tools in helping Japanese language learners to access and record new, target vocabulary.

Longer-term aims were to determine how different CBDs can help maximise lexical acquisition and to consider whether the assistive reading technology these products contain should be regarded as an efficient aid for second language learning. If this is correct, developing and teaching more efficient uses of such multifunctional

CBDs may help language learners quickly confirm, archive, and print meanings in their first language, as well as rapidly access pronunciation of new target language (TL) terms.

Finally, it was hoped that a comparative study of L2 vocabulary assessment would help to isolate which vocabulary skills and strategies are needed for successful language development, and might more clearly delineate the potential enhancing role of CBDs and translation software in more rapid and effective second language vocabulary development.

2. Technological innovations and vocabulary learning

CBDs come in several varieties (see Fauss, 2001; Loucky, 2001), and a summary is provided in Appendix B. Electronic dictionaries are usually pocket size types. Desktop or laptop computer translation software usually has a powerful search engine able to provide immediate translation of words and phrases for various fields of knowledge in multiple languages. This software uses advanced linguistic analysis to process source documents, automatically creating draft quality translations. This analysis is usually based on inputted corpora or natural use vocabulary databases taken from recorded samples of speech or text. Accuracy and the amount of editing required differs highly among products.

Just as Gilbert and Matsuno (2000) have established a multimedia vocabulary database program in Nagasaki as a pilot ESP program for engineering students, similar shared browser programs or online bilingual dictionaries are starting to increase in number, each with various strengths and weaknesses. Since the development of the WWW, various programs have been based on Jim Breen's (2000) excellent EDICT E to J and J to E Files, begun in 1991, but now including recent Palm Pilot adaptations. Logo Vista uses Korya Eiwa's lexicon, offering LAN and Internet packs since 1998. The full version combines OCR and translation modules, and is available for many languages, as is Textbridge's Typist software.

Brother Tsuyaku is capable of basic whole sentence translation. Korya also gives almost instant translation with bidirectional defining ability for English and Japanese, although Japanese meanings are expressed in *kanji*. It also recognizes either singular or plural forms, and can do phrase translation to some extent.

Flatbed and hand-held scanners with optical character recognition (OCR) capabilities can now "read" text, converting it to text files, and obviating typing. The Quickionary II Reading Pen can translate any words scanned into it almost instantaneously and has text-to-speech functions. Scanned words can be stored for later review, and are defined immediately, giving language learners very rapid access to unfamiliar TL forms and meanings. These pens are available in a monolingual English version, and in bilingual versions for 25 languages including Japanese. Likewise, speech-to-text software can transcribe human speech, though often requiring editing, depending on the program and the speaker.

All these tools can support foreign language learning, especially for learners with access to computer rooms with Internet connections. However, most language students need immediate access to new word meanings. Here, portable CBDs can greatly assist low level learners by offering immediate access and confirmation or correction of guesses about the meaning of new words encountered in context. CBD tools greatly speed up accessing and acquisition of new TL vocabulary and grammar forms, and also serve to improve the "hypothetical guessing game," as reading has been characterised (Goodman, 1967; Laufer, 1997). Some computerised tools, just as BBDs, can be bilingual or unidirectional, offering language translation in one direction, such as from English to Japanese. Those that are bilingualised, however, are bidirectional. The Universal Translator software and website is omnidirectional, offering up to 1,560 language translation pairs. Babelfish online gives simultaneous translations for about ten languages. (Others are listed in Appendix C.)

Simple observations and pilot studies (e.g., Fauss, 2001) show the great potential and effectiveness of these bilingualised tools for language learning, especially in enhancing lexical development. Even common sense is sufficient to convince most students of these tools' comparative speed, ease of use, and efficacy in rapidly accessing and ascertaining correct word meanings from multiple possible definitions. Since most EFL readers are slow and "unsafe at any speed" (Eskey, 1979, p. 74), they need all the help they can get in processing foreign language text. Computerised tools can help in at least two ways: CBDs can speed up the processing of meanings needed to comprehend text, whether read or heard. In addition, using text-to-speech functions with any computerised materials may help students read at a more natural speaking speed.

3. Background to the study

3.1 Previous CBD machine translation studies

There are few published studies in the area of electronic dictionaries. A very comprehensive study, Atkins and Knowles (1990) involved about 1,000 language learners in seven European countries, and showed that a majority, in this case 75%, used bilingual dictionaries. Bejoint and Moulin (1987) showed what seems obvious – that bilingual dictionaries are ideal for quick consultations about new vocabulary. Laufer and Hadar (1997) compared the relative effectiveness of various dictionary types for learners at different proficiency levels, testing two different proficiency groups on comprehension of 15 target words, as well as learners' ability to then use the same words in original sentences. Students were encouraged not to repeat dictionary example sentences. The main findings (p. 189) suggested that "different [types of] dictionaries may be suitable for users with different abilities in dictionary use." More relevant to the typical Japanese context of low vocabulary level, unskilled dictionary users are the study's group results, which showed (p. 193) that for unskilled dictionary users:

> The bilingual dictionary produces the best results on the overall dictionary use. In comprehension alone, the highest score was achieved with the bilingualised dictionary . . . [the] monolingual dictionary produced the worst results, significantly in both comprehension and the overall dictionary use.

New language learners need to learn how to access TL lexical items and phrases using various types of dictionary seeking to use all the entry information available to ascertain meaning. A major finding of Laufer and Hadar's study (1997), supported by

Table 1. Dual Assessment Vocabulary Evaluator: Modified Vocabulary Knowledge Scale for Japanese students.

Know L1 (J) definition A (%) 2 points	Know L2 (E) definition B (%) 3 points	Can use word in sentence C (%) 4 Clear or 5 Perfect	Have heard but not sure D (%) 1 point	Unknown word —no idea E (%) No points	Word token or family	Modified ICU # EAP List
					abandon	1
					abbreviate	2
					abide	3
					ability	4
					abnormal	5
					abolish	6
					abroad	7
					absence	8
					absolute	9
					absorb	1

Date: / /   Circle: T1/T2   Receptive% ___ Productive%___

Note. With productive assessment, for words believed known, students write definitions under columns A & B, writing sentences for C on the back. Each word scores 1-10. A perfect productive score is 10 words x 10 points = 100. Compare with Receptive %.

the author's replication thereof (Loucky, in press c), is that a good bilingualised dictionary may be the most versatile and suitable for all types of language learners. Hunt and Beglar also point out that:

Bilingualised dictionaries may have some advantages over traditional bilingual or monolingual ones, essentially doing both jobs. Whereas bilingual dictionaries usually provide just L1 synonyms, bilingualised dictionaries include L2 definitions and example sentences, as well as L1 synonyms. Bilingualised dictionaries resulted in better comprehension of new words than either bilingual or monolingual dictionaries (Laufer & Hadar, 1997). A further advantage is that all levels of learners can use them: Advanced learners can concentrate on the English part of the entry, and beginners can use the translation. (1998, p. 10)

Laufer and Hadar (1997) concluded that as language learners progress, they rely on different parts of bilingualised dictionaries. Monolingual information becomes increasingly important, first in comprehension, using passive understanding or recognition vocabulary, later in production, using active, expressive vocabulary. Laufer and Hadar (p. 195) state, "Even when the monolingual part of the entry is used to its full potential, as in the case of our good dictionary users, the translations may still be helpful in reassuring and reinforcing the learner's decisions about the meaning and use of new words."

Perry (1997) also gives a good overview of the subject. Of the various digitised CBDs available, however, he limits his discussion to English language dictionaries (ELDs), noting their rarity in Japan due to lack of portability. At that time, none were available for portable CBDs. He wondered how well a learners' portable electronic dictionary (PED), monolingual in Perry's use of the term, would be accepted in Japan or Southeast Asia. Most Japanese students still require simpler bilingual dictionaries (CBDs), since many are not yet over the "minimum threshold level" of vocabulary needed to adequately understand or process monolingual dictionary entry information (Laufer, 1997; Loucky, 1996).

3.2 Japanese student vocabulary levels

Most Japanese college students tested in Japan by the author (Loucky, 1996) process, on average, fewer than 100 words of English text per minute when assessed. Anything that helps these students to read and process at a more natural speed can be both beneficial and motivating. The one caveat, however, is that a reading text must be at a particular student's own appropriate reading level, not at his or her frustration level (Ekwall, 1976; Loucky, 1994b), i.e., having fewer than 5% unknown words. Then, even unknown words can be guessed or processed fast enough that text does not become too frustrating.

Using a CBD, however, fewer words would remain unknown. Nowadays, many Japanese college students seldom bring book type dictionaries to school, or use them only infrequently. Those with portable CBDs seem more motivated, appearing to become more proficient in a shorter time. Those who use CBD tools to help speed up accessing of new word meanings can naturally process language input more rapidly. Such students can, in turn, learn to record, remember, and relate or use new lexical meanings, patterns, and phrases with similar rapidity. This is especially true if their software or portable devices have browser functions, hot keys, or flags for instantaneous multi- or bidirectional translation, or function keys allowing users to record, organize, or print new target words stored. Most functions must be taught explicitly to improve student ability to organise their own learning and use of TL vocabulary.

4. The study

4.1 Goals of the study

Research questions examined in this study include:

1. How can the advanced, high speed functions of CBDs be most effectively applied to language teaching and learning?
2. How do CBDs compare to bilingual book dictionaries in speed, accuracy, and adequacy or completeness of definitions?

3. Do CBDs help enhance learner interest, interaction, and motivation levels, and thus contribute to higher levels of vocabulary retention?

4. Are some types of CBD more effective than others, and if so, which technological features and functions are most helpful in the L2 vocabulary teaching-learning process?

4.2 Dual Assessment Vocabulary Evaluator

For this study, the author first categorised common learning strategies for vocabulary acquisition in a *taxonomy of essential lexical processing steps*, focusing especially on the first three steps: assessing, accessing, and archiving new second language target vocabulary (L2 TV). Then, a Dual Assessment Vocabulary Evaluator (DAVE) was designed and used to confirm whether words were known or unknown by students (Table 1). This instrument starts with background knowledge familiar to language learners, namely L1 translations, before asking for L2 definitions or sentence examples. Thus, linguistic demands and anxiety are minimised.

This is *dual assessment* in the sense that it is designed for separate assessment of receptive recognition and productive use vocabulary, usually done one week apart. The former is assessed using the total percentage of words that each student checks as known. The latter (productive) test is assessed by comparing student self-assessments with teacher scoring of responses. Either a simple percentage of accurate responses for each area can be added up, or else a productive score can be calculated by ranking the total amount of knowledge a student possesses for each word on the above scale from 0-10. Ten target words would score a maximum possible 100%. (To compare this assessment tool with other common vocabulary knowledge scales, see Wesche & Paribakht, 1996; Zimmerman, 1997; or Loucky, in press b.)

4.3 Procedure and participants

This study included a total of forty-three participants at four levels of English proficiency as follows: a) 13 pre-advanced level university engineering students; b) 13 intermediate level university engineering students; c) 9 upper intermediate level English majors at a women's junior college; and finally d) 8 lower intermediate computer students from a vocational electronics junior college. They all took a full or partial standardised reading test, the Gates McGinite Form C, at the beginning of this first school year, to determine their English vocabulary grade levels relative to U.S. native reader norms. 90% of all subjects in sub-samples a, b, & d were male Engineering students.

4.4 Word accessment methods

Two means of word access were used. Firstly, for the two engineering classes, the average time for each student to access and record ten items confirmed as unknown using a DAVE was compared. Secondly, the two other classes compared dictionaries by measuring how many unknown words they could look up and record within ten minutes. Surveys were used to determine students' opinions and degree of satisfaction using these means for accessing new foreign language vocabulary.

In each case, the number of unknown words per group of fifty for each student was recorded, along with the time taken per student to find these definitions of new terms using each assigned tool and method.

At first the DAVE was simply given as a word knowledge checklist, a simplified version of International Christian University's Recommended English for Academic Purposes Vocabulary List (Loucky, 1996, pp. 332-349). Then, for further CBD testing and instruction, words marked as completely unknown (Table 1, column E) were highlighted for each student. To ascertain short-term gains in both passive recognition (PR) and active recall (AR) vocabulary, sets of ten unknown words were then assigned to one of several dictionary treatment groups as follows (see Appendix B for a brief review of the functions of each type):

a. Quickionary Reading Pens with OCR and translation capabilities;
b. Electronic pocket CBDs (some used cellphone access to online dictionaries at i-mode and J-Sky);
c. Computer-based translation software;
d. Traditional English to Japanese BBDs (for benchmark comparison with CBDs).

(Other translation software tried experimentally by small groups included Tsuyaku Brother, DTonic, Translator Korya 98, Mac J Dictionary, System Jisho, and Pocket Transer, all desktop CBD software. Other translation software encountered during research included Goma Yakuse, Mikan, and EastJoy. However, time or software version limitations precluded full testing of these.)

Students used their own PED if they had one, Quickionaries borrowed from the teacher, or laptop computers with Brother Tsuyaku and Translator Korya 98 installed.

Students measured accessing speed, recording words with their Japanese definitions. The engineering students were asked to record their own speed when finding words, and all had time measuring devices (stopwatches and cellphone displays). English and computer majors, on the other hand, did not seem prepared to time themselves with such devices, so the teacher did it.

At only one school did time permit testing student ability to write productive sentences using ten target words, via an unannounced quiz one week later. All ten pre-advanced learners scored 100% on this productive retention test.

5. Results

Results are shown in Table 2, which includes both objective and subjective feedback. Overall, both teacher and students were impressed with the versatility and broad potential of these CBDs, and awed by their near instantaneous accessing abilities.

5.1 Comparing class averages
Pre-advanced engineering students accessed words in the order anticipated: Quickionary was the fastest (5.2 minutes for ten unknown words); Laptop Brother translation software was second, followed by electronic CBDs, cellphone CBDs, and finally traditional book dictionaries, which were the slowest (7.2 minutes). Monolingual L1 dictionaries would be even slower, since it takes more time to process, understand, and record definitions given solely in a foreign language.

In addition, pre-advanced students had an average vocabulary level of 5.18, as determined by the reading Gates McGinite pretest. On the Laufer and Hadar replica-

Table 2. Comparative performance of Japanese college student computerised bilingual dictionary use.

Group (N=43)	A (n=13)	B (n=13)	C (n=9)	D (n=8)
Proficiency level	Pre-Advanced	Intermediate	Upper Int.	Lower Int.
Major	Engineering	Engineering	English	Computers
Task	Average time to access 10 wds	Average time to access 10 wds	Average number of wds in 10 mins	Average number of wds in 10 mins
A. Quickionary	5.2 mins	6.8 mins	10 wds	17.25 wds
B. PED (Electronic/ Phone)	5.9 mins/ 6.25 mins	6.06 mins/ 10.5 mins	8.25 wds	19 wds
C. PC CBD (software name)	5.56 mins (PC Brother's Tsuyaku/Korya 98)	5.83 mins	9 wds (DiTonic, Brother Trans, MacJ Dictionary)	11.14 wds
D. BBD	7.2 mins	6.6 mins	10 wds	16 wds
E. Total words accessed per class	40-50 wds (2 also used cellphones: 6.25 mins ave to access 10 wds	40-50 wds (12 also used cellphones: 10.5 mins ave to access 10 wds.	43.5 wds	61.25 wds
F. Favourite dictionary	BBD: 16.7% PC: 33.3% PED: 33.3% Qy: 25% Phone: 0%	BBD: 31% PC: 19% PED: 37.5% Qy: 6.25% Phone: 0%	CBD:100% (BBD: 75%)	Qy: 50%
G. Class average vocabulary level (native norms)	5.18 (PR 75.84%; AP 77.07%)	3.28 (PR 61.58%; AP 72.74%)	3.5	2.5
H. Dy use habits:	BBD: 31% PED: 8% Phone: 15.4%	BBD: 31% PED: 8% Phone: 23%	Only 1 regularly uses PED	3 regularly use PED; 2/3 Chinese used E-J Dy
I. Preferred as possible gift	BBD: 0% PC: 25% PED: 25% Qy: 58% Phone: 0%	BBD: 0% PC: 19% PED: 31% Qy: 44% Phone: 0%	Qy: 100%	Qy: 78%
J. Best for children?	BBD: 75% PC: 17% PED: 8.3% Qy: 0% Phone: 0%	BBD: 37.5% PC: 62.5% PED: 37.5% Qy: 25% Phone: 0%		

Note. PR = Passive recognition on receptive self-reporting Dual Assessment Vocabulary Evaluator; AP = Active productive ability shown on same evaluator; Qy = Quickionary; Dy = Dictionary; CBD = Computerised bilingual dictionary; BBD = Bilingual book dictionary; PED = Portable electronic dictionary.

tion study, the average PR vocabulary level was 76%, and AP vocabulary level was 77%. This is noteworthy, given the very low frequencies of words, and students' limited (30 min) exposure to them. Thus, high level students at or above a fifth grade level, or 5,000 word vocabulary in Laufer's (1997) definition, seem capable of learning from various dictionary types and applying that new learning to comprehension and production tasks.

Intermediate level engineering students also performed close to expectation, except that their use of BBDs came in third place. Laptop Brother translation software was their fastest mode for accessing ten unknown words (5.83 mins). Electronic CBDs were second, BBDs third, then Quickionaries. Finally cellphone CBDs were slowest, taking an average of 10.5 minutes to access ten words.

Of the four levels tested, as expected, higher proficiency students could process new FL vocabulary faster using CBDs than lower proficiency students, for all devices tried. Higher proficiency students tested over two years now have consistently accessed and retained words with increased speed and accuracy, showing CBDs to be of greater value as language proficiency levels increase. Since students with higher levels of language proficiency have been shown to benefit more from monolingual definitions than lower proficiency students (Laufer & Hadar, 1997), it is reasonable to assume that they would also gain more from the use of both CBDs and computerised monolingual dictionaries than less proficient learners would.

5.2 Other findings from the study

Comparing and assessing various types of dictionary this way provides interesting findings. Clearly, monolingual dictionaries alone do not fit the needs of most EFL learners in Japan, since many college students are below the so-called "threshold level" (Laufer, 1997) necessary for their effective use. The rapid access immediate feedback available when using CBDs, and their ability to be bidirectional is of great benefit to foreign language learners for speeding up vocabulary acquisition.

The number of words accessed and recorded by many classes working for only one 40-60 minute session using CBDs was substantial. Most of these 43 students expressed surprise at the sizable volume of new L2 vocabulary accessed, recorded, and at least passively recognized as to initial meaning when they were exposed to four different kinds of CBDs. Lower intermediate level classes averaged 61.25 words per hour, or about one word per minute. If this rate were maintained weekly throughout the school year, 1,800 words could be learned, roughly equal to the total number of all English words required over all six years of secondary school. Upper intermediate classes averaged 43.5 words per class. Pre-advanced and intermediate classes both averaged 40 words (50 if also using cellphone CBDs). The former averaged just 23.9 minutes and the latter 25.3 minutes to access 40 preassigned unknown words.

Finally, this study and other informal surveys often showed that only about 20% of the students regularly carried book dictionaries. Many complained about their weight or slowness, noting that two books would be needed to access vocabulary in both directions. Almost no students carry both. Category F asked students about which CBD or book dictionary they preferred using. Category H asked them to discuss dictionary use habits, and in Categories I-K, students said which CBD they would prefer as a gift, which they thought would be best for Japanese children to use for learning English, and finally gave reasons for their preferences.

5.3 Recommendations for further research
For future study, the next logical step is to compare vocabulary learning rates using various CBD tools. After replicating Laufer and Hadar's (1997) comparison of dictionary types, the author began to investigate what kinds of CBD or translation software could offer Japanese students greater language learning benefits. These might include better learning rates, faster speed of access, greater assistance in accuracy of comprehension and pronunciation, increased learner satisfaction with ease of use, and complete enough meanings to be adequate for understanding in various literary and situational contexts. So far, studies with students of three different majors at several Japanese colleges suggest that Quickionary Pens using assistive reading technology may best meet these conditions. Depending on pricing and better marketing, demonstrations, and sales strategies, they could become increasingly popular, especially in light of positive student response – a high percentage in each class said they would most like to receive a Quickionary of all those CBDs tested.

6. Conclusion

This initial research into the comparative benefits of using computerised dictionaries to increase both speed and accuracy of FL word decoding and vocabulary acquisition helps point the way to the kinds of CBDs needed for better lexical learning. It also helps delineate where teachers and CALL/L2 vocabulary researchers need more rigorous scientific analysis of such computer-assisted learning on an ongoing basis.

Regular use of CBDs appears to help stimulate vocabulary learning. CBDs can help all students process new expressions very rapidly and effectively. Students' ability to expand both receptive and productive language abilities does seem to depend on frequency of use, quality of instruction, and the learner's own initial level of language proficiency, particularly on one's level of known and used L2 vocabulary, meaning those words a learner can not only understand and recognise receptively, but also use productively in their own actual expressions.

CBDs have been shown helpful for the Japanese context. However, students need much more guidance in maximising CBDs' effectiveness by learning to use their various functions at each stage of processing new words using specific essential lexical processing strategies. Moreover, students need not just theory or discussion of why English levels compare poorly with those in other Asian countries, but a practical plan of action using the wealth of modern technology available to this highly developed country. Teachers must help students select the right CBDs (portable, software, or online) to help them expand and maximise their EFL learning. Issues include user friendliness, cost, and computer room scheduling for maximum guided Internet accessibility. Teachers must also show language learners how to select and use relevant information to solve a particular problem or task. In Laufer and Hadar's words:

> Bilingualised dictionaries are a step in the right direction. They provide more information than monolingual or bilingual dictionaries and allow the user to choose explanations in the one language with which he or she is more comfortable, or in both languages for reassurance and reinforcement. (1997, p. 196)

Studies on the beneficial use of CBDs, vocabulary knowledge scales including the DAVE described in this study, and the semantic field keyword approach (Crow. 1986; Quigley, 1986) should be more actively and effectively applied in Japan, which may provide a positive direction for progress, made possible if schools, teachers, and students can gain a vision of how to mine the rich potential of CBDs and well-established vocabulary learning principles (See e.g., Hatch & Brown, 1995; Nation, 2001; and Loucky, 1996, 1997a/b, 1998). In the meantime, the principles of effective vocabulary learning summarised in the taxonomy of lexical processing strategies (shown in Appendix A) should be taught and practiced regularly to help language learners increase their TL semantic knowledge and fluency.

References

Atkins, B. T., & Knowles, F. F. (1990). Interim report on the EURALEX/AILA research project into dictionary use. In I. Magay & J. Zigany (Eds.), *BudaLEX 88 proceedings* (pp. 391-392). Budapest: Academai Kiado.

Bejoint, H. B., & Moulin, A. (1987). The place of the dictionary in an EFL program. In A. Cowie (Ed.), *The dictionary and the language learner* (pp. 381-392). Tubingen: Niemeyer.

Breen, J. (2000). *A WWW Japanese dictionary. (The EDICT Project Abstract).* Retrieved February 2, 2000, from www.csse.monash.edu.au/~jwb/wwwjdic.html

Brother Tsuyaku & Korya 98 translation software [Computer software]. Available from www.systemsoft.co.jp/jiten

Crow, J. (1986). *The keyword approach: Vocabulary for advanced reading comprehension.* Englewood Cliffs, NJ: Prentice-Hall Regents.

Ekwall, E. E. (1976). *Diagnosis and remediation of the disabled reader.* Boston: Allyn and Bacon.

Fauss, R. (2001). *Bilingual electronic dictionaries: Choosing the best* Presentation notes and handouts at PAC 3 at JALT 2001, Kitakyushu, Japan.

Gilbert, R., & Matsuno, R. (2000). Computer-based vocabulary resource database for JSL/EFL education. *The Language Teacher, 24*(9), 15-19.

Goodman. K. S. (1967). Reading: A psycholinguistic guessing game. *Journal of the Reading Specialist, 6,* 126-135.

Hatch, E., & Brown, C. (1995). *Vocabulary, semantics, and language education.* Cambridge: CUP.

Hunt, A., & Beglar, D. (1998). Current research and practice in teaching vocabulary. *The Language Teacher, 23*(1), 7-25.

Laufer, B. (1997). The lexical plight in second language reading. In J. Coady & T. Huckin (Eds.), *Second language vocabulary acquisition* (pp. 20-34). Cambridge: CUP.

Laufer, B., & Hadar, L. (1997). Assessing the effectiveness of monolingual, bilingual, and "bilingualised" dictionaries in the comprehension and production of new words. *Modern Language Journal, 81,* 189-196.

Loucky, J. P. (1994). Teaching and testing English reading skills of Japanese college students. *KASELE Kiyo, 22,* 29-34.

Loucky, J. P. (1996). *Developing and testing vocabulary training methods and materials for Japanese college students studying English as a foreign language.* Unpublished doctoral dissertation, Pensacola Christian College, Pensacola, FL.

Loucky, J. P. (1997a). Maximizing vocabulary acquisition: Recommendations for improving English vocabulary learning for foreign language learners. *KASELE Kiyo, 25,* 101-111.

Loucky, J. P. (1997b). Summary of "Developing and testing vocabulary training methods and materials for Japanese college students studying English as a foreign language." *Annual Review of English Learning and Teaching, 2, 15-36.*

Loucky, J. P. (1998). Suggestions for improving ESL/EFL vocabulary instruction. *Seinan Jogakuin Kiyo, 45.*

Loucky, J. P. (in press a). Comparing and using computerized bilingual dictionaries in Japan. In *JALT & PAC 3 National Conference Proceedings.* Tokyo: JALT.

Loucky, J. P. (in press b). *Designing and testing a Dual Assessment Vocabulary Evaluator.*

Loucky, J. P. (in press c). *Replication of Laufer and Hadar's comparison of dictionary types.*

Perry, B. (1997). Electronic learners' dictionaries: An overview. In P. N. D. Lewis (Ed.), *CALL: Basics and beyond* (pp. 47-50). Nagoya: Chubu Nihon Kyouiku Bunkakai.

Quickionary. (n.d.). *Wizcom Technologies Limited. Assistive reading pen that scans, translates and pronounces.* Available from www.wizcompjapan.com.

Quigley, R. (1986). *A semantic field approach to passive vocabulary acquisition for advanced second language learners.* Unpublished master's thesis, North Texas State University.

Universal Translator. (n.d.). *Orange, CA: Language force.* Available from GoToWorld.com

Wesche, M., & Paribakht, T. S., (1996). Assessing vocabulary knowledge: Depth vs breadth. *Canadian Modern Language Review, 53*(1), 13-40.

Zimmerman, C. (1997). Do reading and interactive vocabulary instruction make a difference? An empirical study. *TESOL Quarterly, 31*(1), 121-40.

Appendix A

Vocabulary learning checklist: Applying taxonomy of vocabulary learning steps, skills and strategies

1) Assessing (Pretest)	2) Accessing	3) Archiving	4) Analyzing
Assessing vocabulary level by VKScales, headwords, or standard test	Meaning-focussed Accessing Definitions: L1/L2; L1 & L2 (Rapid access & recall)	Record definitions with means to recall/study (rapid recording is best)	Rootword-centered word analysis of base, affixes/suffixes
Use EAP VKS sample	"Bilingual is best"	Quickionary OCR/CBD	Word origins/grammar

5) Associating	6) Activating	7) Anchoring	8) Reassessing, reviewing, and recycling
By Semantic Field Keyword Approach (Categorizing by related classes by keywords)	Use focussed new words/phrases activated by productive, expressive use	In short term memory until fixed in long-term memory. Use mnemonic devices.	Measure vocabulary growth/change by #1 Posttest

Appendix B

Comparative computerised bilingual dictionary chart

CBD Name/Type	Cost/Benefits	Features/Functions:	Advantages/Applications for Language Study
Learnout & Houspie's Power Translator PRO7	$100 for English, Japanese, & 5 European languages	Translate full sentences	Multilingual use interface; seamless integration
Learnout & Houspie Easy Language 61 ("See, hear, say it")	$150 for 61 languages; pronounces 125,000 words/21,000 phrases	Voiceprint technology interactive; essential vocabulary focus	Interactive learning games; multimedia video photo tours
Universal Translator	$250 for full suite	40 languages; text to speech & self-read voice translation	Seamless integration; spellcheck 35 languages; poor quality for Japanese
Brother Tsuyaku	¥10,000	Full phrase translation ability	Excellent quality for reasonable price
Ditonic (East Joy)	About $50; very rapid access	Uni-directionality, only E to J	Enhanced grammar study of various word forms
MacJ Dictionary	Shareware	Rapid access English, Japanese, Korean	Rapid search functions
System Jisho (Mac)	System shareware	Clear & user-friendly	Bi-directional searches
Pocket Transer: Mac/PC	¥20,000	Full translation; save & print functions	Excellent rapid access; archiving & printing
Atlas, V7	$450	Only E to J	

Note. Prices and exchange rates change. Current prices are listed for comparison only.

CBD Name/Type	Cost/Benefits	Features/Functions:	Advantages/Applications for Language Study
Unico Sura Sura Series	Separate OCR & translation packs	6 languages available for OCR/translation	
Honkaku Tsuyaku (Sourcenext)	¥5,500; add-on tech. lexicons	Bidirectional with voice translation	
Soiku Hayawaza JET	¥14-20,000 2 versions	Choice of six tech. vocabulary areas; add-on ESP lexicons extra	High-speed translation (Ad: "Translates 14 pages in 29 seconds!")
Tsuyaku Ichiban (Souiku)	¥8,000	Uses Pocket Transer translation engine as database	
Tsuyaku no Tetsujin	¥3,500	Uses Pocket Transer as database (ASCII)	Disadvantage: EJ & JE sold separately
Tsuyaku Pikaichi, V3	¥7,000	Bi-directional; works in Office XP/2000, Word	Translates PowerPoint
Tsuyaku Office	¥6,000	Translates ME/95/98 or NT4/2000 documents	
Tsuyaku Ohsama, V4	¥6,000	Internet/E-biz Aim: for all platforms/Java	Works with Linux, Lotus Notes, etc.
Translator Korya 98 Ippatsu Tsuyaku & Loga Vista X	$200 Logo Vista OCR/ $100 Translator only. 8 available languages.	Reads text to speech, translates spoken or written input	Translates most MS Office & 2000 documents and websites/email
Quickionary Reading Pens	$200-250	OCR handheld scanner Very fast	L1 translation/L2 pronunciation capabilities
Eijiro	$400-500	Very fast	With sound
Ippatsu Tsuyaku	Includes some Fujitsu		
Rakuchin	Fujitsu system	Bi-directional	With sound
Bookshelf	Shareware; needs CD to use	Several databases	Word, grammar, people, proverbs
Cannon 300 PED	¥7,000	3 archiving functions	Idioms, related phrases
EZ-Word PED	¥4,000	No archiving	
Seiko II PED	¥3,000		Limited English meanings
Sony DD-S25	Slower, easier to damage, most expensive	Primarily for specialists (translators)	Voice function has fairly good quality; of limited value
Casio XD-S3000	Many useful learning aids not provided by a standard dictionary	Longman is a particularly helpful addition	Good for students serious about English study
Casio XD-1500; Seiko 7200 & 8000			All promising for English study

Appendix C

Comparing online computer based dictionaries

Website name	URL	Features/ Functions	Advantages/ Applications
ALIS Gist in Time	www.alis.com/cgi-bin/transdemo.pl	10 languages available	Free service
InterTran	www.tranexp.com:2000/InterTran		Free service
Language Engineering Corporation	www.lec.com/demo/frame.html		Purchasing fee
Amikai Web Translation Engine	www.amikai.com/	Used by various sites. Can do 24 language pairs including URLs!	Free service; excellent quality for news, etc.; printable quality
AltaVista Babel Fish Translation	world.altavista.com/tr	Systran Internet world keyboard	Free text or website translation of 12 languages
Excite Japan	www.excite.co.jp/world	Uses Amikai Engine	
@nifty Global	www.02.so-net.ne.jp/~suyama/	Uses Amikai Engine	
Cafeglobe	Cafeglobe.com/café/wotg/index.html	Website translation	Requires membership
Rikai.com (Good for news)	rikai.com/cgi-bin/HomePage.pl? Language=Ja	Single word choices	Trilingual reading development (archives)

10

Student Self-Evaluation Skills Using CHILDES Programs

Peter John Wanner
Kyoto Institute of Technology

1. Introduction

This preliminary study introduces a powerful method of transcribing oral conversation that makes use of cognitive skills while learning a second or foreign language. Before the appearance of computer programs that could analyse quantitative features of language development, transcribers relied heavily upon manual methods for determining productive language development such as vocabulary lists, mean length utterances, maximum length utterances, time duration speaking, and turn taking. Likewise, Japanese students enrolled in EFL classes at Kyoto Institute of Technology (KIT) initially used only manual methods for analysing conversations. However, preliminary findings indicate that after using computers to record transcripts and automatically analyse conversations using the Child Language Data Exchange System (CHILDES), student fluency increases.

CHILDES is a collection of computerized transcripts of typed and handwritten examples of natural dialogue of children (MacWhinney, 1995). Its principles of transcription use the acronym CHAT (Codes for the Human Analysis of Transcripts). The transcript analysis program is known as CLAN (Computer Language Analysis). For the past twenty years, researchers have extensively used CHILDES and other similar programs for analysing child language. Trying to teach these programs to students with the aim of improving their speaking ability was not recommended. However, this paper will explain and discuss how college students learn and apply the programs within CHILDES to develop further awareness, and help increase communicative fluency.

2. Method

Audio recording cannot preserve as much conversational detail as video. The larger the group size, the more important it is to have video to capture these intricate interactions. With pairs, voice recognition is sufficient to distinguish who is speaking during each turn interval. With six students, however, it is difficult to identify with voice recognition alone, but video recordings show clearly who is speaking.

Recording facilities should be in a room that can amplify the voices of individual speakers. If the room is fairly small, the maximum group size would be six to eight people because the video camera needs a certain distance to include everyone in the frame. Group learner self-evaluation video (GLSEV) has shown positive signs of language development in a preliminary study of first year college students when implemented with manual CHAT transcription and CLAN analysis (Wanner, in press). When these programs are applied to smaller groups, such as pairs, audio recording alone might be sufficient. Murphy et al. (1998) used video recordings for pairs and had students evaluate their discussion using a more general and less quantitative means of analysis, learner self-evaluation video (LSEV). Students examined their conversation and filled out a survey, rating themselves on certain areas from 1 to 10. The increased subjectivity compared with GLSEV can make it difficult for students to evaluate and follow their progress.

This paper explains GLSEV and basic applications from the CHILDES programs in a preliminary study of 51 first-year students at KIT, a national university in Japan. All had technology-related majors, and were taking a communication course in EFL. Students were required to discuss a prearranged topic each week; they were prohibited from bringing books or materials into the discussion room. Their topic was known well in advance and assigned as homework to prepare for the next week. Students were assigned to groups of six, and preliminary analysis of transcripts throughout the semester (13 sessions) provided evidence of communicative development.

2.1 Facilities and background

Every week, students left the classroom for 20 minutes to discuss their theme on video. They were assigned to permanent groups with letter assignments in each discussion room so they could transcribe the group interactions effectively. While three groups of students were in discussion rooms, the other three groups were in the classroom with the teacher; they then exchanged places. Each discussion room had one video camera.

Although most first year students at KIT have some access to computers before college, their knowledge is extremely limited; only 30 percent of the class were computer literate and owned a computer. It was necessary to provide a computer room and teach the other 70 percent basic computer operation before showing them how to use the CHILDES programs, which took a considerable amount of time outside class. Sixty-five students required approximately four hours to demonstrate both basic computing and CHILDES. Some students were so intrigued with the CHILDES program to analyse their productive development that they bought a new computer shortly after enrolling in the class. Computer literate students needed little help since the program was largely self-explanatory.

2.2 Recording procedure
Students turned in VHS tapes for dubbing, collected tapes for evaluation, turned in 3.5 inch floppy disks with transcript analysis, and collected floppy disks with transcript analysis during a four week period following the first class. Students had two VHS tapes; one was placed in a box each week to be dubbed. The tape recorded from the previous week was taken home. Hence, one VHS tape and floppy disk were for odd numbered thematic discussions and another set for even ones. In the second week, students put the even tape in a box for lesson 2 on that day and took out the odd VHS from the box for the previous week's lesson. They then had one week to transcribe and analyze their first lesson conversation using the computer. The third week they put in the odd VHS tape set at the end of their last conversation two weeks prior (L1), picked up the conversation tape from the past week, and turned in a disk for transcription and analysis for lesson 1. This procedure continued throughout the semester.

Since the teacher dubbed tapes for students each week following their discussion and returned them the following week, it was necessary to have at least six or seven VHS recorders to reduce dubbing time. The teacher also checked disks and videos simultaneously to ensure that students had transcribed correctly.

2.3 Codes of Human Analysis for Transcripts (CHAT)
2.3.1 Principles
The CHAT system is a standardized format for computerized transcripts of face to face conversational interaction between any number of interlocutors (MacWhinney, 1995). All data within the CHILDES system is in CHAT format. Transcripts in this format are designed to facilitate the subsequent analysis of transcripts by the CLAN programs outlined later. It is designed to give a precise and uniform description of conversational interactions between interlocutors. Therefore, it is important to record word or word group repetitions, pauses, interruptions, mispronunciation of words, and other forms of speech. Students transcribing must initially learn basic coding options and include file headers providing information related to the context of the interaction. They must also understand how transcriptions are done on the "main line," as this is where they record their main conversational interactions.

2.3.2 Obligatory headers
Two file headers are necessary to transcribe data. The first type is *obligatory headers*, without which the CLAN program will not operate. On the first line of each week's transcription, students type "@*Begin*." On the second line students add "@*Participants:*" followed by a three-letter code (i.e., STA, STB) and their interlocutor role (i.e., Student). The end of the transcription is indicated by "@*End*." Table 1 shows a sample transcription with these two entries for a group of six interlocutors.

2.3.3 Alternate headers
Table 2 provides examples of headers not required for the program to operate, but which provide important information. These are known as *constant headers* and

changeable headers. First, students enter the conversation date following the header "@Date:" (i.e., 1-JUN-2002). Second, they add the theme after "@*Comment:*" (e.g., Country). Then, they type the colour of their group after "@ group of #:" (i.e., Orange). Fourth, their identification number is inserted after "@*ID:*" (i.e. 102500500bL1=STB). This consists of the student's school ID, followed by a capital letter representing their colour group, a lower case letter representing their assignment in the group, the lesson number, an equals sign and "ST," and finally a capital letter for their assignment in the group. The only "changeable header" students use is to keep track of the length of time they speak in each conversation, "@Time Dura-

Table 1. Initial obligatory transcription example.

@Begin
@Participants: STA Student, STB Student, STC Student, STD Student,
STE Student, STF Student
@End

Table 2. Constant and changeable header example.

@Date: 7-MAY-2001
@Comment: Country
@Group of #: Orange
@ID: 012500500bL3=STB
@Time Duration: 00:14-1:00

tion:" (e.g., 12:30-13:10).
One transcription symbol all students must use is "*STB:" which shows it is student B's speech that they are entering. If the interlocutor changes, the letter following "*ST" changes (see Appendix A). Students are not required to type out the whole conversation as this would be too demanding; instead, they only type what the interlocutors before and after them say.

2.4 Transcription
When transcribing, students type all words including Japanese and English in their conversation. English words use orthographic spelling except where there is a slightly different pronunciation when spoken. In the following example the word "off" is pronounced as "offu":

    *STB: offu(off) shite (verbal phrase).

In the above example, closed parentheses follow a Japanese word or English word with nonstandard pronunciation. Students must translate all Japanese words that they or their preceding or following interlocutor use. Furthermore, all sentences begin with lowercase letters except when the first word is a pronoun or proper noun. This is designed to help the CLAN program distinguish between regular words and proper nouns when producing frequency counts, and will be discussed in more detail below. When students cannot hear or understand another student, they can use *xxx* or *xx* strings as follows:

    *STA: xxx.
    *STB: what?
    *STA: I would like to xx.

If students cannot make out the pronunciation, they can use *yyy* or *yy* strings:

    *STA: yyy
    %pho: /bukuabui/
    *STB: what?
    *STA: yy I yy.
    %pho: /buku/ /ai/ /bui/

The CLAN program will include xx or yy but not xxx or yyy strings in frequency counts. Repetition of words or phrases is recorded using a slash mark and the number of repeats inside closed parentheses following the word or utterance. Below is an example of a one-word utterance repeated once followed by another utterance repeated three times:

    *STA: I [/] think I would go to California.
    *STA: it's very [/3] warm there.

If an utterance of more than two words is repeated, caret marks "< >" are required. If the multiple utterance is repeated more than once, the slash "/ x times" is followed by the number of occurrences inside parentheses, e.g.,

    *STB: <I think> [/] I would go to Washington.
    *STC: <I would like to> [/2] go to Alaska.

Sometimes students spell out words when their interlocutors cannot understand their pronunciation. In this case, the CLAN program allows two ways for transcribing letters so that each one is not counted as a word. The first is to spell out the word with an at-mark (@) representing each letter, e.g.,

    *STA: what did you say?
    *STC: I said have you ever had a 'b@l 'a@l 'r@l 'b@l 'e@l 'q@l 'u@l 'e@l

The second way is to use the corresponding consonants and for the vowels say the actual sound. For example "*ay*" would represent the letter A and "*ai*" would represent the letter I, e.g.,

    *STA: what did you say?
    *STC: I said have you ever had a b ay r b e q u e

Acronyms are another common feature of conversation. Proper names like All Nippon Airlines would be "A+N+A" or nicknames like "J+J" for Jessie. Video cassette recorder becomes "v+c+r" and is not capitalized but uses the plus sign. Numbers such as 22 must be written out as "twenty+two" and times like 11:00 am must be written out as "eleven o'clock ay m." Titles such as Mr. or Dr. must be written out as in "Mister Smith" or "Doctor Yamada."

The CLAN program requires a termination marker at the end of every sentence. The period, question mark, and exclamation mark can occur only once and must be in sentence final position. Furthermore, only one sentence can exist per transcript line.

Quoting is another frequent occurrence in conversation. When a quote follows what is said, it is written on the next line with quotation marks:

> *STA: and then my mother said +"/.
> *STA: +" no, you can't go.

If a quote precedes what is said, it is written on the preceding line as follows:

> *STA: +"no, you can't go.
> *STA: my mother said +".

Students mark pauses with a hash (#) sign. For 0-5 seconds students use one hash sign, 6-10 seconds, two hashes, and 11 to 15 seconds, three. The following is an example of significant pauses within a sentence:

> *STA: being a doctor is ah, #, difficult job.
> *STA: they have to um, ## work long hours ### and they don't get many vacations.

When calculating times from duration recordings, students subtract the number of seconds they speak. If a student is silent for longer than 15 seconds within a single sentence, they write the time of silence following a "####" sign as follows:

> STB: traveling by train is, um, ####(:35) very crowded.

The symbols above are the most important for transcribing data in CHAT format. Of course, there are many others that help provide more detailed analysis such as actions and overlap. However, many of these other features are not generally evident in Japanese students' English conversations; Appendix A provides evidence that pauses are more prevalent.

2.5 Transcription check
Before a student uses the CLAN program, they are strongly advised to check their transcript. In most cases the CLAN program will not operate unless this is done successfully. Appendix B, describes the procedure to use the *check* function. Following this, the student is ready to start analysing the document.

2.6 Computerized Language Analysis (CLAN)

2.6.1 Setting the working directory

The CLAN program can do a large number of automatic analyses, but for EFL classes, three areas are most appropriate for analysing basic conversational fluency development. To run any of these programs, students go to the column in the tools window and scroll down to the command row. A box appears with four boxes as follows:

> *Working c:\CHILDES\CLAN\lib*
> Output
> Lib c:\CHILDES\CLAN\lib\
> mor lib c:\CHILDES\CLAN\lib\

If the file is saved in the library of the CLAN program within CHILDES, the file is accessible and automatic CLAN analysis is possible. However, if trying to access a file from the desktop, temporary, or any other file, it is necessary to reset the working file. Once this file is selected, the desired programs can be accessed.

2.6.2 Frequency

The program most often and easily used constructs a frequency word count for type and token words, as well as a type-token ration (the total number of word types divided by the number of occurrences in the specified file). Having opened the command box and set the working file, double clicking the CLAN icon on the bottom left of the command box will give an alphabetical list of commands available. Choosing the command *freq* brings up the following in the display box:

> freq

Typing "+*f*" creates a new file with the frequency list for the file being analysed. Adding a space followed by " ı *t*" instructs the computer to look for only frequencies of the specific interlocutor following it, using the "*" plus the interlocutor's three letter code (e.g., STB) plus a space, i.e.,

> freq +f +t*STB

Clicking *File in box* and selecting a file to analyse from the *File to choose from* box (e.g., 0125005OOb.L3.cha) will bring this file to the right hand side of the "please select files" box highlighted. An at-mark symbol now appears:

> *Freq +f +t*STB @*

Finally, pressing *done* should produce a message similar to the following:

> From file <c:\temp\01240050RbL3.cha> to
> file <c:\temp\01240050RbL3.frq.cex>
> Done with file <c:\temp\01240056ReL3.frq.cex>

If the message *Done with file* does not appear, then some error has occurred and the computer will usually identify which type. Appendix C is an example of a successful frequency count analysis with a complete list of word types and tokens for each word for the CHAT transcription file in Appendix A. At the bottom of Appendix C is the following analysis output:

> 71 Total number of different word types used
> 118 Total number of words (tokens)
> *0.602 Type/Token ratio*

Students must further break the word type and token counts into the number of words used for each language because the computer cannot distinguish between them without a special coding system. This is very cumbersome to learn considering the short time frame, so students quickly scan the frequency counts and count Japanese word types and tokens. Then, they subtract these from the total word types and tokens. For Student B in this transcript, all words are English and the following would be recorded in the yearly recording master sheet:

> Percent English 100%
> *Percent Japanese 0%*

2.6.3 Mean length utterance (MLU)
The second type of CLAN analysis students do is mean length utterance. Having opened the command box and set the working file, they choose the command "mlu." Procedure is identical to that for *freq* above. The following should appear in the dialogue box for file 10240050RbL3.mlu.cex:

> mlu +f +t*STB
> Wed Aug 01 22:55:17 2001
> mlu (23-Jan-2001) is conducting analyses on:
> ONLY speaker main tiers matching: *STB;
> **
> From file <c:\temp\0125005OOb.L3.cha> to
> file <c:\temp\0125005OOb.L3.mlu.cex>
> MLU for Speaker: *STB:
> MLU (xxx and yyy are EXCLUDED from the utterance and morpheme counts):
> Number of: utterances = 34, morphemes = 118
> Ratio of morphemes over utterances = 3.471
> Standard deviation = 4.024

The most important information for students is the number of utterances or sentences, because they can follow progress in these terms. The morpheme counts, ratio of morphemes over utterances, and standard deviation are less important and students pay little attention to these when analysing progress. However, these numbers are also included on the master recording sheet.

2.6.4 Mean length turn (MLT)
The third type of CLAN analysis is a MLT analysis. Again, a similar procedure to that

above is used, except that "mlt" is initially selected. Below is an example result for file 10240050RbL3.mtu.cex:

mlt +f +t*STB
Thu Aug 02 00:18:18 2001
mlt (23-Jan-2001) is conducting analyses on:
ONLY speaker main tiers matching: *STB;
****
From file <c:\temp\01250050Ob.L3.cha> to file
<c:\temp\01250050Ob.L3.mt.cex>
MLT for Speaker: *STB:
MLT (xxx and yyy are included in the utterance and morpheme counts):
Number of: utterances = 34, turns = 23, words = 118
Ratio of words over turns = 5.130

MLT is important because it provides the students with information on how interactive they are in the conversation group. Usually more turns indicate more interactiveness, given satisfactory MLUs instead of short one or two word confirmations or constant questions. If students only take one turn of 118 words, the ratio would be 118. In most cases, at least an average number of around 5.0 words per turn is preferable. If a student speaks too much during one turn, this might be considered overbearing. These numbers are also recorded on the student's yearly master sheet.

2.6.5 Time duration
The final analysis is to record the total speaking time or total duration. Students must check the CHAT transcript manually and add up the seconds they spoke, subtracting pauses from the time duration totals for each utterance. Transformation into minutes and seconds is done manually because the CLAN program is not yet programmed for this. Students normally manage to scroll through their 20 minute transcript of utterances fairly quickly. Time duration is also added to the student's yearly master sheet.

Table 3. Yearly recording master sheet.

Theme:		5/24	
Word Type	71	Turns	23
Word Token	118	Words	118
Type Token Ratio	.602	Ration of Words/Turns	.530
Japanese Word Type	0	Ratio of Utter./Turns	1.478
English Word Type	71	Ratio of Words/Utter.	3.471
Japanese Token Type	0	Time Duration	1:12
English Token Type	118		
Percentage Japanese Type	0		
Percentage English Type	100		
Number of Utterances	34		
Morphemes	118		
Ratio of Morphemes/Utter.	3.471		
Standard Deviation	4.024		

2.7 Yearly recording master sheet

2.7.1 Self-evaluation applications

Table 3 shows a section of a yearly recording master sheet providing students with a summary of figures for all the four areas they looked at.

2.7.2 Self-implementation incentive

This sheet can help students try harder to improve quantitative scores for spoken fluency based on the four areas discussed above, since they can monitor progress constantly and always know where they stand in the class. Students can use this tool for assigning themselves grades at the end of the term. If all students followed the guidelines for analysing data honestly, teachers could work more as facilitators teaching vocabulary and culture as well as helping students use the program. However, this is not the case in this preliminary study of Japanese students taking conversation classes with this method of self-evaluation. Sometimes, students tried to abuse the system, thinking the teacher would not notice the following:

- Copying and pasting one conversation over the next three or four discussions;
- Not recording Japanese utterances that they or other interlocutors speak;
- Altering quantitative figures on the yearly recording master sheet;
- Substituting another interlocutor's conversation transcript.

This provided disincentive to other, hard-working students because those doing very little work receive a similar grade. If this occurs, the only way to provide a fair grading system is for the teacher or assistant to check transcripts and video-tapes for the above abuses. In this study, five percent of students used one of the above abuses when recording original transcripts from video-tape. Hence, the teacher checked all transcripts against the video recordings to verify students were transcribing correctly.

3. Conclusion

Appendix D shows a summary of three lessons over six weeks for one student on a small section of a yearly recording summary sheet. The CLAN frequency results of this student for Lessons 5, 6, and 9, over approximately four to six weeks, provide evidence that the percentage of English word types used in conversations increased and furthermore the percentage of English word tokens also increased, while Japanese decreased. In addition, the number of utterances did not vary much between the two middle weeks (L5 & L6), but towards the end of the term (L9), the number of utterances increased by over 40 percent. Furthermore, the ratio of morphemes over utterances decreased over 32 percent. Finally, the subject took more turns in the final conversation. Thus, the ratio of words to turns for lesson 9 is less than that in lesson 5 and lesson 6. The ratio of utterance over turns for lesson 9 is equal to lesson 5 and less than lesson 6. Finally, the ratio of words over utterances for lesson 9 is less than lessons 5 and 6. Hence, preliminary findings would tend to indicate that use of more conscious processes to develop language through the GLSEV and CHILDES programs helped students to produce more English vocabulary with less Japanese in one semester. A more detailed analysis of a larger sampling will shed more light on these findings.

Note

The software on floppy disk or CD-ROM can be ordered via the following address:
Lawrence Erlbaum Associates
365 Broadway, Hillsdale, New Jersey 07642, USA
Tel: 1-800-9BOOKS9; Fax: (202) 666-2394
Email: ORDERS@LEAHQ.MHS.COMPUSERVE.COM

The program is also available for free download from:
<childes.psy.cmu.edu> <jchat.sccs.chukyo-u.ac.jp/JCHAT/>
The second website provides installation instructions in Japanese. Once installed, students can start transcribing their transcripts. The CHILDES programs can be installed on DOS, Macintosh, and Windows computers.

References

MacWhinney, B. (1995). *The Childes Project: Tools for analysing talk* (2nd ed.). Hillsdale, New Jersey: Lawrence Erlbaum.

Murphey, T., & Kenny, T. (1998). Intensifying practice and noticing through videoing conversations for self evaluation. *JALT Journal, 20*(1), 126-140.

Wanner, P. (in press). Implementing Group Learner Self Evaluation Video. *Memoirs of the Faculty of Engineering and Design, 52.* Kyoto Institute of Technology.

Appendix A

CHAT transcript example
@Begin
@Participants: STA Student, STB Student, STC Student, STD Student, STE Student
@Date: 7-MAY-2001
@Comment: Country
@Group of #: Orange
@ID: 012500500bL3=STB
@Time Duration: 0:08-0:12
*STB: would you like to move a new country?
*STB: which one?
@Time Duration: 0:16-0:22
*STA: yes, I[/2] want to move to Spain.
@Time Duration: 0:31-0:32
*STB: why?
@Time Duration: 0:37-0:42
*STA: my favorite architect , # architecture.
@Time Duration: 0:47-0:47
*STA: gaudy!
@Time Duration: 0:48-0:48
*STB: gaudy?

@Time Duration: 0:48-1:04
*STA: gaudi , go , I want to see his buildings , # XXX Sagurada familia.
*STA: do you know?
@Time Duration: 1:05-1:05
*STB: yes!
@Time Duration: 1:06-1:06
*STA: yes?
@Time Duration: 1:08-1:12
*STB: yes, I do.
*STB: yes, I do.
@Time Duration: 1:15-1:18
*STA: so [/2], I want to go to Spain.
@Time Duration: 1:26-1:27
*STC: what would you miss most?
@Time Duration: 1:46-1:47
*STC: what would you miss most?
@Time Duration: 2:01-2:02
*STB: nothing?
@Time Duration: 2:06-2:10
*STA: if I go to Spain?
@Time Duration: 2:13-2:13
*STB: yes!
@Time Duration: 2:16-2:23
*STA: maybe , # my family , Only!
@Time Duration: 2:26-2:26
*STB: your family?
@Time Duration: 2:27-2:30
*STA: my family and my girl friend!
@Time Duration: 2:32-2:35
*STB: oh ,# who?
@Time Duration: 2:38-2:38
*STA: secret!
@Time Duration: 2:45-2:55
*STA: then ,where your, ## where have you want to move?
@Time Duration: 2:56-2:59
*STB: no [/3].
@Time Duration: 2:59-3:00
*STD: question?
@Time Duration: 3:00-3:01
*STB: question [/2]!
@Time Duration: 3:03-3:13
*STD: what would your biggest problem be?
@Time Duration: 3:46-4:10
*STA: umm, ## in this college I can study Spanish , So if I want to go, I study by myself.
*STA: that's problem.
@Time Duration: 4:13-4:14
*STB: okay.

@Time Duration: 4:31-4:32
*STB: nothing.
@Time Duration: 4:41-4:52
*STA: then you ,# where [/2] do you want to move?
@Time Duration: 4:54-4:55
*STB: country?
@Time Duration: 4:55-4:56
*STA: yes.
@Time Duration: 4:56-6:01
*STB: I [/2] want to go to Canada [/2].
*STB: because [/2] I'm studying English and French now , so , # English and French
is spoken in Canada.
*STB: so , I can , I can , communication them.
@Time Duration: 6:06-6:09
*STD: what do you miss most if you go to Canada?
@Time Duration: 6:10-6:26
*STB: umm , # my friend , I miss friend.
@Time Duration: 6:30-6:30
*STD: who?
@Time Duration: 6:32-6:42
*STB: junior high school , high school , college , student , friend.
@Time Duration: 6:55-7:52
*STB: problem [/2]?
*STB: my problem.
*STB: if I go to Canada , problem is , # my problem is , # in fact , I'm not good at studying
French.
*STB: I [/4] will study in French very hard , very [/2] hard.
@Time Duration: 8:18-8:21
*STD: what would you like to move a new country?
@Time Duration: 10:05-11:10
*STD: yes , e , # I want move to America , famous America.
*STD: because I live in Siga in Japan , but Siga is country.
*STD: so I want to go, # .
@Time Duration: 11:11-11:12
*STB: city?
*STB: big city?
@Time Duration: 11:13-11:14
*STD: big [/2] city!
@Time Duration: 14:27-14:47
*STD: I want to study abroad someday , so I , # , I work arbite (part+time+work) to
go to America.
@Time Duration: 14:49-14:50
*STB: really?
@Time Duration: 15:00-15:04
*STE: what job will you do?
@Time Duration: 15:00-15:25
*STD: I cooked spagetty and pizza and coffee.

*STD: irassyai (Hello)!
@Time Duration: 15:28-15:29
*STB: where?
*STB: guest?
@Time Duration: 15:30-15:31
*STD: guest?
@Time Duration: 15:31-15:33
*STB: Skylark?
@Time Duration: 15:38-15:44
*STD: not popular , # , not famous.
@Time Duration: 15:47-15:48
*STB: aha , not famous!
@Time Duration: 15:49-15:52
*STD: local [/2] kissate(coffee).
*STD: nante (what) iuno (say)?
@Time Duration:15:53-15:55
*STB: coffee, local coffee!
@Time Duration:15:56-16:04
*STD: local coffee!
*STD: but , delicious.
@Time Duration: 17:44-17:46
*STC: do you like hamburger?
@Time Duration: 17:47-17:48
*STE: yes!
@Time Duration: 17:49-17:52
*STB: I like big mac.
@Time Duration: 17:56-18:02
*STA: eighteen yen hamburger , I like.
@End

Appendix B

Automatic check performance function

To perform an automatic check, go to the "Mode" column in the tools window at the top and scroll down to "Check opened file" or press escape-l. The computer will automatically go through the transcript and check all entries. If something is entered incorrectly, the computer will mark and flash at the location of the error and provide an explanation at the bottom. For example, the following message might appear:

CLAN [E] [CHAT]*9 Utterance delimiter expected.

Furthermore, line 9 of the transcription would have the space following "country" at the end of the sentence flashing to indicate the error location:

*STB: would you like to move a new country

In this case, a period needs to be added after country because some type of punctuation – a period, or question or exclamation mark – is necessary to show the end of the sentence. The program can check intensively for spelling and grammar or for basic formatting only. When a successful check is completed, the following message appears:

CLAN [E] [CHAT]*: Success. No errors found.

Appendix C

Frequency count of Appendix A
freq +f +t*STB
    Wed Aug 01 22:31:52 2001
    freq (23-Jan-2001) is conducting analyses on:
    ONLY speaker main tiers matching: *STB;
    ****

From file <c:\temp\01250050Ob.L3.cha> to
file <c:\temp\01250050Ob.L3.frq.cex>

1 a	1 family	1 miss	1 study
1 aha	1 famous	1 move	2 studying
2 and	4 french	3 my	1 them
1 at	3 friend	1 new	4 to
1 because2	1 gaudy	1 no3	1 uhm
1 big	2 go	2 not	1 very
1 big+mac	1 good	2 nothing	1 very2
2 can	1 gucst	1 now	1 want
2 canada	2 hard	1 oh	1 where
1 canada2	2 high	1 okay	1 which
2 city	9 I6	1 one	1 who
2 coffee	2 I'm	5 problem4	1 why
1 college	1 if	1 really	1 will
1 communication	3 in	2 school	1 would
2 country	3 is	1 skylark	4 yes
2 do	1 junior	2 so	1 you
2 English	2 like	1 spoken	1 your
1 fact	1 local	1 student	

 71 Total number of different word types used
118 Total number of words (tokens)
0.602 Type/Token ratio

Appendix D

Sample student summary

Student name	01250037Ya		
Start date	5/21/01		
End date	7/02/01		
	L5	L6	L9
Theme	5/21	5/28	7/2
Word type	50	33	32
Word token	66	61	74
Japanese word type	11	5	4
English word type	39	28	28
Japanese word token	11	5	4
English word token	55	62	70
Percentage Japanese type	22%	15%	13%
Percentage English type	78%	85%	87%
Type/token ratio	.758	.541	.432

11

Security on the Internet: Resources for Teachers

James Duggan
Dokkyo University

1. Introduction

The ability to access the Internet from a personal computer has enriched the lives of many households in Japan and around the world by providing new sources of information, entertainment, and communication. Online news, entertainment, and email have at least partially replaced newspapers, television, and postal mail for many. The Internet has also done much for teachers and students, both in the classroom and out. Connecting to the Internet has allowed both teachers and students to improve the learning experience by giving them access to a variety of useful teaching and learning resources (e.g., Duggan, 1998b, 2000, 2001).

The realization of the value of the Internet by the public, government offices and business corporations, and teachers and school administrators has led to an increasing demand for Internet connectivity. This is turn has required an expansion in networking – an increase in the number and thickness of connections carrying the data – resulting in increasing use of fibre optics in data cable and advances in technology to allow faster connections over existing copper phone lines. The average user sees this expansion in highspeed Internet access in the replacement of common phone line access by cable modems and digital subscriber lines (DSLs). In Japan, where cable modem and DSL access is increasing, highspeed access is fast becoming the connection of choice, replacing not just common phone line access, but ISDN as well.

There may, however, be a price to pay for an increasingly accessible Internet. With the global surge in Internet use, there has been a corresponding surge in computer/ Internet-related crimes and security risks – including such threats as viruses and personal intrusions. These risks are echoed by many concerned with computer security (e.g., Ricciardi 1999, p. 247; Machrone, 2000, p. 101; Morris, 2000, p. 74; Seltzer, 2000, p. 62; Baker, 2001; Gibson, 2001b; Sweet, 2001, p. 44).

Most users, lacking any such personal experiences, probably feel that such security risks on the Internet happen only to other people. Events in the news such as the I Love You and Melissa viruses have received much coverage in the international press, as have the hackings or attempted hackings into various government and corporate computers. Identity theft and credit card fraud are increasingly in the news as well (Zone Labs Inc., 2001a). But why should teachers be concerned?

The fact is that anyone, whether living and working in Tokyo or New York, who accesses the Internet is at risk. Granted, the reader may not have encountered any threats up to now, but the chances of having one's security – or that of one's students and institution – compromised are strongly increasing. There is even a very good chance it has already happened, because unless without knowing where to look and what to look for, one might never notice. This potential risk from Internet access requires teachers to be especially aware and vigilant and demands a very real need to understand how to deal with such risks. Like any individual, teachers need to protect themselves, but as teachers, they have an even greater responsibility in protecting students and institutions.

The purpose of this paper is a practical one – to help teachers be better prepared when using the Internet by demonstrating and explaining a variety of risks involved (the *threats*). As the threats are varied, and because no single product addresses all the security threats posed by the Internet, this paper suggests some specific defensive (and offensive) measures, including how to deal with viruses, worms, scans, and probes (the *solutions*).

First, a few items concerning this paper need to be pointed out:
• The solutions given in this paper are aimed especially at those with limited knowledge of how computers work. There are, or course, much more advanced or technical solutions, but those contained herein are meant to be achievable for the average teacher.
• This paper concerns, in general, free or low-cost solutions since classroom budgets are limited (Duggan, 1998a). Again, many good options are also available for a price, but this paper's suggestions should be more than adequate.
• As the author is most familiar with Windows PCs, this paper will be Windows-centric. Corresponding Mac, and often UNIX, versions are usually available, and can often be found at the sources given in this paper.
• While the specific solutions in this paper are available for, and have been tested on, English operating systems, this author has successfully installed the recommended personal firewalls on Japanese OS computers. Japanese versions or programs, such as antivirus software, may also be available. The threats are the same regardless of OS.

2. Computer viruses

2.1 The threat

A teacher opens email and finds a message waiting from a friend, containing an attachment named "Resume.doc." Is he perhaps recommending a colleague for a job? Well, though the school is not looking for anyone just now, it would not hurt to have a look at the resume. Or would it?

Opening the attachment has released the now-infamous "Killer Resume" virus onto the teacher's computer, where it proceeds to erase all data and files from all drives. Because Microsoft Outlook is the preferred email program, the virus is also sending copies of itself to everyone in the address book, where the cycle will repeat itself, destroying files of all friends, family, and acquaintances.

Does this sound plausible? To protect one's system against such a scenario, one should first better understand computer viruses by determining what computer viruses are and how they work, as well as an understanding of the tools and techniques used to fight them.

What is a computer virus?

Simply put, it is just a small computer program. Like any other, it contains instructions telling a computer what to do. Unlike an application, however, a virus usually tells the computer to do something unwanted, such as damaging or destroying data.

Computer viruses can be thought of as the "common cold" of modern technology. The computer virus attaches itself to other programs just as a human virus would attach itself to cells, thus infecting them. Also, like a human virus, computer viruses replicate by themselves and spread to other hosts. They can quickly move across open networks such as the Internet, causing expensive damage in a relatively short time. Around 1995, according to Zetter (2000), the chance of receiving a virus over a one-year period was about 1:1,000; more recently, this has increased to about 1 in 10.

How does one get a computer virus?

When viruses first appeared in the 1980s, they spread very slowly, usually through trading and sharing floppy disks. The introduction and popularity of the Internet has provided viruses with a much more efficient way of spreading. Especially, the ability of computer viruses to target email programs is now considered the greatest contributor to their rapid spread, accounting for about 81 percent of infections (Zetter, 2000). Well over 500 million emails were sent and received every day back in early 1999 when the Melissa macro virus struck (Mitchell, 1999). It is certainly more today.

What can a virus do to a computer?

Virus behaviour can range from annoying (such as causing a computer to make random bleeps, or flashing a message) to destructive (reformatting a hard drive or deleting necessary system files). According to Mitchell (1999), about 30 percent of viruses are malicious.

How a virus specifically affects a computer depends on the type of virus infecting it. The three most common types are the *macro* virus, the *file infector* virus, and the *boot sector* virus.

Macro viruses target the documents of specific software programs based on the macro programming languages. Common examples of such programs are the Microsoft Office applications Word and Excel. From these documents, macros infect the template on which the documents are based, and then infect every document with the same template. Because each macro virus is specific to a type of application, it cannot infect a different application. For example, a Word macro virus would not infect an Excel spreadsheet.

Because of the popularity of programs such as Word and Excel that use macro language and the fact that macro viruses can be spread through nearly any medium (diskettes and CD-ROMs, email attachments, downloaded files, file transfers), macro viruses are the most common active virus type presently reported by users. It is also the fastest-growing class of virus (Schurman, 2000). One of the best-known is Melissa, which caused nearly US$100 million in damages when released in early 1999. Melissa spread by attaching itself to Word documents sent as email attachments.

File infector viruses target executable files. When an infected application is opened, it activates the virus, spreading it to other executable files on the hard drive that are also open or running. File infector viruses can be transmitted in a variety of ways: through files on diskettes and CD-ROMs, as email attachments, and from Internet downloads.

Boot sector viruses are so called because they replace or implant themselves in the section of a hard drive or diskette drive known as the boot sector, that part that tells the PC how to load the operating system. If infected, the user may not be able to "boot up" (start) the computer. These viruses spread by copying themselves from an infected computer to the boot sector of a diskette, and from there to the boot sector of the hard drive of the next PC that uses the infected diskette.

Though similar to viruses in that they are harmful programming, *worms* and *Trojan horses* are not true viruses. Like a virus, a worm will replicate itself, but unlike a virus, which requires some action on the part of a user to spread to other computers, a worm will move copies of itself to other computers without outside help. Like the "I Love You" virus, which was actually a worm, worms do not have to wait for the user to send an email; they do it themselves. Unlike true viruses, worms also do not alter or delete files, instead often replicating inside the host computer, consuming storage space, eventually causing the overloaded computer to slow to a crawl.

Unlike worms and viruses, Trojan horses do not replicate and spread themselves, but are stealthily inserted into the target computer by an intruder. The Trojan horse is used to either gather information from or control of the host computer. As the threat and style of attack of a Trojan horse is different from that of worms and viruses, they will be addressed separately, later in this paper.

A greater threat than sending a virus to others on a mailing list is that of bringing one into one's own institution's network. What should concern teachers and students is that a single virus, worm, or Trojan horse placed onto one computer within an institution through carelessness can do untold damage.

2.2 The solution: Antivirus program

There are a number of precautions to take; the first is to get a reliable antivirus program. Antivirus experts agree that diligent use of a reliable antivirus program is the best method of protecting a computer from virus-related problems. This software protects a system by examining (scanning) files (and sometimes emails) for the signature (a unique string of bytes that identifies the virus like a fingerprint) of a known virus or other program, and then removing the virus. Two of the best known and supported are Norton AntiVirus and McAfee VirusScan.

Norton AntiVirus must be purchased, but is often bundled with a new computer, and is part of the Norton SystemWorks package. A free trial version can be downloaded from the Symantec website (www.symantec.com).

McAfee VirusScan must also be purchased, but a free demonstration program can be downloaded from Download.com (www.download.cnet.com).

InoculateIt Personal Edition is available free from Download.com. Though this author has never personally used this program, it is put out by a reputable company (Computer Associates), is recommended as the best antivirus freeware program by PC Magazine (Grubbs, 2000) and the Home PC Firewall Guide (n.d.), and is one of the top downloads at Download.com.

An important criterion for selecting an antivirus program is whether it has regular and easily available virus definition updates. Antivirus programs can only identify and remove known threats. As there are always new viruses coming out (and spreading quickly), it is important to keep "signature" files updated by downloading the latest updates from to maintain the software's effectiveness. Experts estimate that 300 new viruses are created each month (Mitchell, 1999). Virus definition updates for Norton AntiVirus, McAfee VirusScan, and InoculateIt Personal Edition can be downloaded free from Download.com.

For added safety:

- Avoid opening unexpected email attachments and downloads from untrustworthy sources. If an unexpected file attachment arrives, contact the sender before opening it. Ask if they meant to send the file, what it is, and what it should do. Special care is needed if the attachment has an .EXE extension (indicating an executable program), or a .VBS extension (which the infamous I Love You worm has taught users worldwide to be wary of).
- Regularly back up files and data. Not only might a virus corrupt some or all files, but the hard disk may someday fail, or system hacked into and data destroyed.
- Download the latest patches or versions of programs to fix any security holes found or exploited and regularly check the manufacturer's website for security alerts.
- Be aware that commonly used applications, such as Microsoft Word, Excel, and Outlook, are more often targeted by viruses and their creators. While discontinuing the use of such software would do much towards eliminating the threat to a system, this is infeasible. However, if suspect Word documents are received by email, open them in Wordpad rather than Word, since macros will not be understood by Wordpad. Also, if continuing to use Microsoft's Outlook software, Machrone (2000, p. 101) suggests disabling its automation features to prevent mailing lists being hijacked.

3. Intrusions (Hacking)

3.1 The threat

You are online, perhaps browsing the headlines for topics to teach, purchasing books at an online bookstore, or checking email from students. But at the same time, unknown to you, someone is scanning the computer for a way to get in, or may already be searching through its files, stealing whatever is of interest or value, and destroying the rest. Excerpt from a sci-fi novel? The plot of a new movie?

What few teachers consciously consider is that access to the Internet goes both ways: When connected to the Internet, not only can you connect to other computers, but they can also connect to you. A highspeed broadband connection such as a cable modem or DSL is especially vulnerable, a point repeatedly stressed in the recent literature (e.g., Machrone, 1999a, p. 81; Ricciardi, 1999, p. 247; Thomas, 1999; Morris, 2000, p. 74; Seltzer, 2000, p. 62; Sengstack, 2000b; Denton, 2001, p. 49; and Gibson, 2001b).

Broadband connections are increasingly popular because of their performance and convenience benefits, greatly surpassing those of a dial-up connection. Besides the obvious speed advantage (download and upload), connections are permanently on, and so do not require users to connect each they want to use the Internet. This very advantage is what makes broadband connections more susceptible to attack.

Systems with broadband connections are typically targeted because they are the simplest to track down. Whenever a computer logs on to the Internet, it has an Internet protocol (IP) address assigned to it by the Internet service provider (ISP) used. This IP identifies the computer to the network. Without an IP address, the sites you connect to would not know where to send the information the computer has requested (such as a page download). A small utility such as IP Agent [available for free download from Gibson Research (grc.com)] will show the present IP address.

Computers using an "always on" broadband connection (cable, DSL) are typically assigned fixed or *static* IP addresses. Dial-up modem connections typically "borrow" an IP from the ISP while online connected to the Internet. Upon disconnection, the IP is returned; this is termed a *dynamic* IP address.

Regardless of the kind of connection (static or dynamic), the longer that address is "active" or online, the better the chance an outsider can find it. The difference is that the static IP used with a broadband connection is often left active even when the user is not online; even if the user breaks the connection, the IP address will be identical the next time the user logs on.

A hacker determines who to attack through what Scambray et al. (2001, p. xxvii) refer to as "target acquisition and information gathering": First, he runs a scan of a range of IP addresses, usually of known cable or DSL addresses. After this scan has run its course, he goes over the scan logs, looking for openings, servers, and hints as to files.

Thus, a hacker can always find someone with the static IP address of a broadband connection. With a dial-up connection, if the hacker does not get a user immediately, by the time he has gone over the logs and picked the person out, the user is probably already offline, only to go online again at a later time, but with a different IP. Though dial-up users with a dynamic IP address are not completely safe from intruders, they do face less risk. The true risk rests with broadband connections. So anyone with a static IP address and poor security is vulnerable to attacks from the Internet – not only when online – but 24 hours a day.

So why would a hacker want to hack a teacher's or school machine, especially since it is not a government or big corporate server? After all, the Internet is a big place, and with so many computers out there, who would notice one little computer? Besides, who would want to hack this computer specifically – it is not a mission-critical corporate system holding vital company information.

This belief is outdated. While it is true the stereotypical *hackers* of quasi-movie-hero-like fame would have very little interest in one personal PC, it is not these people we should be mainly concerned with. The *crackers* and *script kiddies* are the ones who want to get into one's system, and it is to these that Machrone (1999a, p. 81) refers when he points out that "the number of people with moderate hacking or cracking skills – and tools – has increased dramatically."

The term hacker is generally used to describe anyone who breaks into a computer, for any reason. Members of the hacking community, who only hack to prove it can be done or to expose a security weakness, like to make the distinction between themselves and criminal or malicious hackers, referred to as crackers.

Crackers can be classified into four categories: those wanting information, such as credit card numbers; those wanting to destroy data; those wanting to alter data, such as grade fixing; and distributed denial-of-service attacks, which will be addressed later in this paper. The motivation for such attacks ranges from financial to revenge to peer respect (Jones, 1998).

The third term, script kiddie, is used by the hacking community to describe those first-time hackers who make use of easily available scanners and hacking techniques with little knowledge of what they are doing, or understanding of the damage caused.

Even as the number of crackers increases greatly, Machrone (1999a, p. 81) explains that the real population explosion has been among the clueless millions of newcomers to the connected world. These users are a field day for malicious crackers who leave their scanners running night and day, looking for easy or interesting targets. Gibson (2001b) tells of one hacker who states that he is able to scan 30,000 sockets per minute, and can even "scan some small countries in one night."

The increasing number and sophistication of hackers makes any Internet-connected PC a potential target. Crackers understand that virtually every Internet-connected PC can hold valuable information – financial statements, bank account and credit card information, and personal details.

One cracker nicknamed Anthrax (n.d.) boasted recently on a hacker website that he has "gained everything from games to credit card numbers stored in Quicken databases to music videos. The possibilities are endless!" he crows. This cracker has also conveniently included directions on how to begin hacking immediately and has the recommended program and password cracker ready for immediate download (for Windows PCs). This author has tested the information and programs presented, and found them a credible threat.

Even if a cracker was not interested in data, he might be interested in one's computer. He may just wish to hide behind it when attempting to hack into other systems, or use it in a nasty new kind of denial-of-service attack – the distributed denial-of-service attack (DDoS). A DDoS attack targets a specific system, attempting to swamp its defences and bring the network to its knees by flooding it with useless traffic from multiple commandeered computers. This has recently brought down a number of systems, including those of EBay, E*Trade, CNN, and Yahoo. Gibson (2001c) tells in detail of a string of DDos attacks that shut down his security website by a 13 year-old who commandeered 474 computers from around the world for the attacks.

The sheer number of scans run on the Internet would indicate that the reader most probably has already been scanned. According to International Data Corporation, the average broadband connection experiences three attempted intrusions in its first 48 hours (Sonicwall, 2001, p. 82). U.S. news reported on a reader's computer that had 538 attempted intrusions over a two-month period – an average of almost nine every day (Thomas, 1999). This author assumes that these reported instances are not the more serious hack, which is a determined specific attempt to compromise one computer, but a simple scan or probe looking for anything vulnerable. This author usually finds himself the target of such scans or probes an average of once every 5 to 6 hours online (See Fig.1).

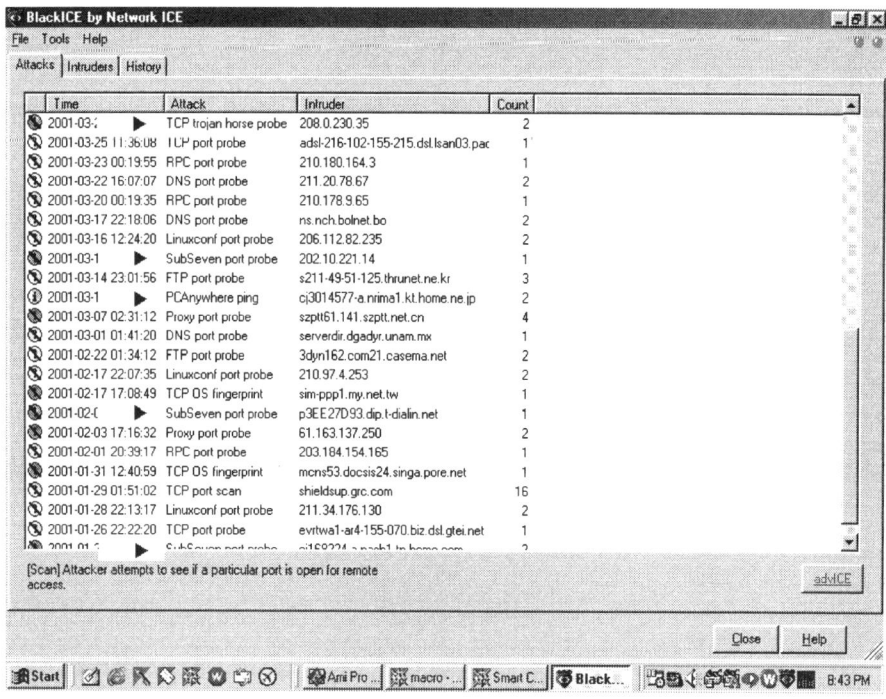

Figure 1. Page shot of attempted intrusions on the author's PC, as logged by BlackICE Defender.

Scans and probes on a system do not show up because, when not running a server, attacks are not logged unless using a firewall that logs them. So, while probable that a system has been scanned or probed, it is much less likely – though possible – to have been compromised. The chances are it was ignored because the used a dial-up connection until recently, and looked singularly uninteresting. The computer probably runs Windows and has unremarkable names for user, computer, and workgroup names (such as the default "Workgroup"), information easily obtainable and logged by a cracker. Unless a computer is named "Swiss accounts," a cracker would probably go looking for something more interesting.

This does not, however, mean no attacks in the future. An individual's vulnerability increases with the adoption of a broadband connection, greater numbers of crackers, technological advancement of hacking tools and programs, and the adoption of security techniques by other users. While the frequency of security risks through intrusions may never equal that of viruses, incidents of computer hacking have risen (Sweet, 2001).

Given the damage that can occur though hacking, it is important to take all possible safety precautions.

3.2 The solution: Personal firewall

The traditional approach to Internet security and the single best way to prevent unauthorized intrusions is with a *firewall*. As the name suggests, this acts as a barrier, blocking out intrusions by inspecting every packet of data as it reaches the computer and determining whether it should be blocked. Gibson (2001a) describes in more detail how firewalls work at "Personal Internet firewalls that really work!" (grc.com/su-firewalls.htm). Two basic types are available: hardware and software (personal) firewalls.

Hardware firewalls are more appropriate for SOHO (Small Office/Home Office) LAN (Local Area Network) security. Because of the cost and difficulty in configuring a hardware firewall, the best solution for the individual user may be a personal firewall. Like a hardware firewall, this filters and selects which packets of information are admitted and which are blocked, based on user specified rules.

Personal firewalls are necessary to:
- properly and safely monitor Internet connection for intrusion attempts;
- allow remote access of computer files across the Internet, or through any sort of Internet-based remote control or access program such as PC Anywhere;
- pre-emptively protect from compromise by "inside the wall" Trojan horse programs like SubSeven and Back Orifice.

Today, a number of personal firewall products are on the market. [Home PC Firewall Guide (www.firewallguide.com), provides a detailed analysis of firewalls and available choices.] Among these, a special type of personal firewall product – intrusion detection programs – has quickly gained popularity. Two often reviewed and recommended are Network ICE Corp's BlackICE Defender and ZoneLab's ZoneAlarm (Ashworth, 2000; Sengstack, 2000a; Sheldon, 2000; Strauch, 2000; Boran, 2001).

Intrusion detection programs can be thought of as the computer equivalent of a burglar alarm. They watch all Internet protocol transmissions to and from a PC and identify and block any malicious intrusions. These programs also collect information on these attempts (see Fig. 1).

An intruder will scan, probe, or attempt to access a system through an available port. In general, applications have default port numbers. For example FTP (File Transfer

Protocol – the method used on the Internet for sending files), SMTP (Simple Mail Transport Protocol – a common method of retrieving email), and POP3 (Post Office Protocol 3 – a common method for sending email) all use default port numbers, and are targeted by intruders.

Because such a program can detect and block incoming intrusions to these and other ports, your computer will appear nonexistent to the cracker. Checking a firewall – or vulnerability due to lack of one – can be done using the ShieldsUP tests at Gibson Research (grc.com). Running these tests from behind an institution's firewall is not advised without first consulting the System Administrator, as this will be seen as a direct attack on the institution's system. The tests can, however, safely be run the tests on a computer at home.

Unlike many other firewall programs, both BlackICE and ZoneAlarm are easy to set up and use. Neither requires sophisticated programming of rules for operation, nor knowledge of TCP/IP port or packet types. Both BlackICE and ZoneAlarm provide real-time security feedback and alerts and a choice of security level; BlackICE offers four levels of protection – Trusting, Cautious, Nervous, or Paranoid. ZoneAlarm allows different settings for LAN and Internet connections.

The major differences between BlackICE and ZoneAlarm are:
- BlackICE 2.9 can be bought via direct download from NetworkICE's website <www.networkice.com>, while ZoneAlarm 2.6 is free for personal use <www.zonelabs.com>.
- BlackICE collects much more precise and detailed information on the intruder, such as IP address and DNS name and identifies the attempt as a specific intrusion type (as shown in Fig. 1) than does ZoneAlarm.
- BlackICE blocks only incoming attacks. ZoneAlarm also stops attacks that start from within a PC (such as Trojan horses and other unauthorized programs that attempt to secretly transfer personal information from a system over the Internet).

Is a personal firewall needed behind a school firewall? While probably well-protected from outside hacking threats when using a computer on an institution's network and behind its firewall, in practice, teachers and students must be extra-cautious. If a hacker does gain access to the network through teacher or student computers, the security and data of the entire network is at risk. Network security can be compromised unwittingly by, for example, downloading a Trojan horse, or connecting to the Internet through a common modem while still linked to the network.

Even if a network is behind a firewall, Machrone (1999b, p. 81) warns that others on the same network behind this firewall can easily hack in. The institutional firewall offers no defence. Thus, anyone with File and Print Sharing turned on allows others free access to their printer and hard disk contents. An intrusion detection program should block them from snooping around, although one will need to enter a list of "trusted IP addresses" or face a barrage of false alarms.

Are all attacks really attacks? Looking at Figure 1, the three Trojan horse probes (black arrows) outlined earlier are malicious and might be a threat if one had inadvertently picked up the Trojan being scanned for. The PCAnywhere ping is most likely an innocent mistake. The other scans/port probes, while not an immediate threat, may constitute a future threat if a firewall is not used.

Firewalls, including those presented in this paper, tend to overwhelm the user with data. Firewall "newbies" may be stunned at how often their PC is scanned. Much contact, however, is innocent traffic (false positives) due to an Internet service such as a chat, FTP, or multimedia server (video, audio) sometimes sending data to the wrong IP address. A firewall might interpret such activity as a scan.

However, given the number of intended intrusions mentioned above, some scans and probes may be intentional. According to Network ICE, a security software provider and the creator of BlackICE Defender, port probes are the most common intrusion detected on the Internet. The typical hacker may scan thousands or millions of machines in a single session. "There are always thousands of script-kiddies out there, so if you have an always on connection (cable-modem, DSL), then you can expect about one of these (TCP port probe) scans per day." (Network ICE, undated c)

Given that this author has logged several Trojan horse probes within a set time frame (Fig.1), common sense indicates that one could expect the intentional scanning of other ports as well, as there are a variety of probes and scans at least as common as Trojan horse scans (Network ICE, undated a-d). So, while not every scan is deliberate, and few, if any, will result in an attack, one would be wise to defend against the possibility of such an attack.

A personal firewall should stop would-be crackers and script-kiddies, who do not have the ability or resources to overcome a firewall. However, a firewall is at best a deterrent, because a determined hacker could probably break through it, just as a professional thief could steal a car or break into a home regardless of any precautions taken. Even so, it will deter a hacker, since so many other users take no security precautions at all.

4. File and print sharing

4.1 The threat
The single greatest security hole that exists, and a commonly abused one, is Windows File and Print Sharing over a TCP/IP (Internet) connection. McClure and Scambray (1998) and Rubenking (2000, p. 208) point out that many freely available scanners and password crackers are used by crackers looking specifically for computers running Windows File and Print Sharing. What is mesmerizing to crackers about accessing another's computer through Windows File and Print Sharing is its ease and ability to see every file on the computer, including the "C" drive, just as if looking at "My Computer" on your own PC. One can then exam, alter, copy, or delete the contents to one's heart's delight:

> Leaving File and Print Sharing turned on in the Network section of Control Panel for your Internet connection . . . is rather like leaving your door open and placing your wealth in full view a few feet inside the door. Password-protecting your shares . . . is like placing a speed bump on the road. It'll slow the criminals down, but it's just a matter of time before they get you. (Ricciardi, 1999, p. 247)

4.2 The solution: Disable file and print sharing

An Internet-connected computer not networked to any other machines has absolutely no need for file and printer sharing. You can disable it manually, or through a personal firewall; both BlackICE and ZoneAlarm allow this easily. Furthermore, ZoneAlarm can allow local network File Sharing access, while disabling Internet File Sharing access.

If, as some educators wish, a teacher who has to share files across the Internet, with students, office, family, or friends, must address some serious security concerns. Not only will user, computer, and workgroup names be public knowledge, but so will the names of all shared resources. In fact, with a cable modem, leaving File and Print Sharing on could cause a computer to pop up as available when someone else on the local network segment opens his or her Network Neighbourhood icon.

To share files between two specific computers, such as at home and work, it is better to create a secure "tunnel" across the Internet which has no danger of unauthorized intrusion. With firewall technology, this is made possible and relatively simple by instructing the firewalls to allow connections from the other computer's IP address. Thus, both machines can "see" the other, but no one else on the Internet knows that either machine has established such a secure tunnel across the Net.

Even Anthrax, the cracker introduced earlier, advises readers to "Never enable (Windows) File and Print Sharing unless you absolutely have to!" (Anthrax, n.d.).

5. Backdoor programs (Trojan Horses)

5.1 The threat

A common way for a cracker to compromise security and gain control of a computer system is through a backdoor program, such as Back Orifice or SubSeven, which installs itself in Trojan horse fashion, arriving as an email attachment or posing as some useful executable program.

Trojans are becoming increasingly popular in data and password theft, where the aim is not mass disruption (like a virus) but usually the silent theft of secret information, though the intruder's purpose may be to install a virus, reformat a hard drive, or other malicious action. The purpose of a particular popular class of Trojans, Remote Access Trojans, is to provide the cracker with complete remote control over a machine.

According to Gibson (2001c), the highly invasive SubSeven Trojan grants the hacker absolute control over commandeered machines. Among the many invasions the SubSeven Trojan enables is monitoring every keystroke to capture online and eBanking passwords and credit card numbers. The SubSeven Trojan has been implicated in the recent DDos attack on the grc.com website.

While one would never install a backdoor program on a system knowingly, it may be inadvertent; Trojan horses can be acquired in a number of ways. One such trap is for someone pretending to be a representative of Microsoft or a service provider sending an email urging someone to download what they claim is a critical software update, when in fact, it is a disguised Trojan horse. Installing it makes one open for attack. These programs wait on the system until contacted by the cracker from his computer, and allow this person to log on to a computer undetected and take over its system completely – reading files, watching keystrokes, or even rebooting the machine.

A similar threat can be had for users of remote access programs such as PC Anywhere. The cracker will probe for the existence of such a program, and then it just remains to crack the password, a security threat we will assess shortly.

For more detailed information on SubSeven and BackOrifice, as well as other Trojan horses, see the *Database of Intrusions Detected* by Network ICE (advice.networkice.com/advice/Intrusions/) and Firewall Guide's Anti-Trojan page (www.firewallguide.com/anti-trojan.htm).

5.2 The solution: Personal firewall/application control

The answer is, of course, to not install any unsure programs or items in the first place. But even if a system were to inadvertently pick up a Trojan horse, through, say, an infected email, a personal firewall will still prevent network access to the backdoor. This is because no passing Trojan scanner could detect or know of the Trojan's existence since all attempts to contact it inside the computer would be blocked by the firewall.

Such Trojan scans are very common, as shown in the author's recent BlackICE logs (Fig.1), where three SubSeven Trojan probes were successfully blocked. The author has also logged a number of Back Orifice scans in the past. The PC Anywhere ping marked is most likely accidental (a false positive), as its IP address is in the author's local segment. If an IP address is from outside the local area, this should be considered a clear attack against the system.

In addition, the kind of application control which is part of the ZoneAlarm personal firewall is highly effective in preventing a Trojan horse or rogue program from sending data out to the Internet. Application control also makes it impossible for a computer to be used in denial-of-service attacks, in which remote hackers send commands to compromised computers to launch attacks.

To help in assessing a system's application control security, LeakTest, a tiny freeware program available at the GRC website (www.grc.com), simulates a Trojan/ spyware program on a system that attempts to slip past personal firewall's defences and connect to an Internet server. If LeakTest fails, the firewall passes. If it succeeds, the system is vulnerable. Finnie (2000), McWilliams (2000), and Machrone (2001) all found that ZoneAlarm passed where other software firewall programs failed.

6. Password crackers

6.1 The threat

A password has great value in protecting secrets. Whether protecting a computer, financial and credit card information, or student grades and exams, a good password is one of the best defences against hackers which is why hackers want them. "Ask ten hackers what the greatest prize is and nine will probably answer 'your passwords.'" (Zone Labs, Inc. 2001b)

Password-cracking programs are abundant, and include crackers for BIOS passwords, Zip files, POP mail, operating-system log-ons, MSN Hotmail, and even brute-force PGP. One website contains more than fifty. One of the most notorious, L0phtCrack, has been downloaded over half a million times. An administrator who tried L0phtCrack was astounded to find "that it uncovered 85 percent of his office's passwords in a mere 20 minutes and that it unlocked all but two in 24 hours" (Machrone, 1999a, p. 81). How safe is yours?

6.2 The solution: Use difficult passwords

Password crackers are typically based upon a dictionary of proper names because most people choose names of children, pets, or relatives as passwords. Since these programs have all the time in the world – one has no idea when they are grinding away trying to get in – this guessing approach usually succeeds eventually.

Therefore, employ cryptic and unguessable passwords. As a cracker has access to computer names, passwords should therefore have no relationship to any of these. Ideally they should be long random strings of characters, for example, "4wE9sd27HRaQ9." While annoying to have such a password, any meaningful phrase might either be in a cracker's dictionary or be guessed by a friend or acquaintance. It is better to reduce an easily remembered phrase such as "My daughter Maia was born on 8/5 in Tokyo" to "MdMwbo85iT."

Having to crack long, mixed-case passwords with special characters can add days or weeks to the task. Change passwords often, and never give them to anyone. Do not enable "Save password" because if someone gains access to the computer (whether over the Internet or in person), they will also have access to whatever is being protected.

7. Taking offensive measures

7.1 Report utilities

It is, initially at least, unnerving to find oneself scanned and probed by unknown crackers. One way to fight back is to notify the intruder's ISP and get their account cancelled, effectively knocking them off the Internet, though possibly just temporarily. Firewall report utilities from Brady & Associates (ClearICE Report Utility 4.0 and ClearZone) integrate into the BlackICE Defender and ZoneAlarm intrusion detection firewall programs, and support and supplement them by organizing all pertinent information on the attacks. They then automatically put this into an email, ready to send off to the intruder's ISP, which BlackICE or ZoneAlarm should have identified.

This author has used this procedure a number of times, sometimes successfully (according to feedback from the intruder's ISP), and sometimes unsuccessfully, as many ISP's have neither time or inclination to pursue the matter. This is understandable, as countless scans and probes occur every day. Given this great frequency, it is not time-efficient to report every scan, but only those that target a system specifically, or where the intruder has a record (according to the intrusion detection program logs) of attempted intrusions. This is especially true when one realizes that, as according to Network ICE (undated c), 10% of such scans are from spoofed or innocent addresses and another 20% from compromised systems.

Both the ClearICE Report Utility 4.0 and the ClearZone Report Utility are available for download from the Brady & Associates website (www.y2kbrady.com/

firewallreporting/).

7.2 Port scanners

Port scanners are, of course, a cracker tool that can be used right back at such an intruder. For the intruder to be intruded upon can be quite a surprise, as most crackers believe their activities to be stealthy.

However, port scanning software can be a defensive as well as an offensive tool. It tests networks to determine which ports are vulnerable to attack, so one can take steps to ensure system security. ShieldsUP, the online diagnostic scanning and probing program available at Gibson Research (grc.com), also tests ports.

There is a variety of Windows-based scanners readily available for download on the Internet. Though none is as robust or flexible as their Unix counterparts, all provide a surface view of security holes in a network. Two freely available scanners are Port Scanner – extremely simple yet effective shareware from Blue Globe Software (www.blueglobe.com/~cliffmcc); and Internet Maniac – a freeware utility by Sumit Birla (members.tripod.com/~Sumit_Birla) that does more than just scan ports.

Note, however, that scanning others' ports is considered rude, and could lead to being reported to the ISP and resultant account closure.

8. Conclusion

When utilizing security resources, as with many utilities, there are often trade-offs. In this case it includes the time and work involved installing antivirus and anti-intrusion programs, updating them, and keeping students aware of the risks involved. Further inconveniences may be encountered when trying to access one's system from the outside or access the Internet through a secure firewall, in addition to handling virus and intrusion alerts. However, very real possibilities of risk arising from Internet access as presented in this paper far outweigh what inconveniences may arise from ensuring a secure environment, not just for one's own files and data, but for the security of students and institution.

Every user is vulnerable to some degree most of the time. Vulnerability levels will depend on the amount of Internet use, the kind of information on the computer, and time and money invested in security. While a determined hacker could probably defeat most security measures, constant vigilance, good habits, and making use of the latest security technologies should put off most intruders most of the time.

It is hoped that by demonstrating and explaining a variety of the risks involved in accessing the Internet, as well as by suggesting some specific defensive (and offensive) measures, teachers will now be better prepared to deal with these risks, and so be able to more safely and securely make use of the teaching and learning resources of the Internet.

References

Anthrax. (n.d.). *NetBIOS hacking through Windows made easy.* Available from warex.box.sk/howto/netbioshack.htm

Ashworth, R. (2000, December 9). Protecting your home computer from the Internet, can you keep the heat out? *SANS Institute.* Available from rr.sans.org/homeoffice/heat.php

Baker, T. (2001). Potential threats to your PC's security. *Smart Computing, 12*(2), 38-43.

Boran, S. (2001, April 26). Personal firewalls/intrusion detection systems: An analysis of mini-firewalls for Windows users. *Security Portal.* Retrieved from securityportal.com/articles/pf_main20001023.html

Denton, C. (2001). Put your PC behind a firewall. *Smart Computing, 12(2),* 49-51.

Duggan, J. (1998a). Technology in education: Utilizing educational software for language learning. *Dokkyo University Studies in Foreign Language Teaching, 17,* 119-164.

Duggan, J. (1998b, May). *Utilizing the Internet as a teacher resource: Traditional and non-traditional uses.* Paper presented at the Institute of Foreign Language Education, Dokkyo University, Japan.

Duggan, J. (2000). An Internet resource: Planning for your next conference. In P. N. D. Lewis (Ed.), *Calling Asia: The proceedings of the 4th Annual JALT CALL SIG Conference* (pp. 69-72). Nagoya: Chubu Nihon Kyouiku Bunkakai.

Duggan, J. (2001, May). *Security and privacy on the Web.* Paper presented at the JALTCALL 2001 International Conference, Kanto Gakucn University, Japan.

Finnie, S. (2000). GRC's LeakTest 1.0. *The broadband report, 1*(36). Retrieved from content.techweb.com/winmag/columns/broadband/2000/36.htm#securityctr

The Firewall Guide. (n.d.). *Firewall guide anti-virus.* Retrieved from www.firewallguide.com/anti-virus.htm

Gibson, S. (2001a). *Personal Internet firewalls that really work!* Available from grc.com/su-firewalls.htm

Gibson, S. (2001b). *Shields UP! Internet connection security for windows users.* Available from grc.com/su-danger.htm

Gibson, S. (2001c). *The strange tale of the denial of service attacks against grc.com.* Retrieved from grc.com/dos/grcdos.htm

Grubbs, L. (2000, May 30). Freeware solutions: Vanquishing viruses. *PC World.* Retrieved from www.pcworld.com/hereshow/article.asp?aid=16916

Jones, C. (1998, February 20). Internet hacking for dummies. *Wired News.* Retrieved from www.wired.com/news/technology/0,1282,10459,00.html

Machrone, B. (1999a). How do I hack thee? *PC Magazine, 18*(21), 81.

Machrone, B. (1999b). The enemy within. *PC Magazine, 18*(22), 81.

Machrone, B. (2000). Always more threats. *PC Magazine, 19*(13), 101.

Machrone, B (2001, January 22). Does your PC leak? *PC Magazine.* Retrieved from www.zdnet.com/pcmag/stories/opinions/0,7802,2671115,00.html

McClure, S., & Scambray, J. (1998, July 6). Free Windows-based scanners are plentiful, but only Asmodeus shows promise. *Security Watch.* Retrieved from www.infoworld.com/cgi-bin/displayNew.pl?/security/980706sw.htm

McWilliams, B. (2000, December 7). Personal firewalls fail the leak test. *InternetNews-Intranet News.* Retrieved from www.internetnews.com/intra-news/article/0,,7_529661,00.html

Mitchell, R. (1999, April 12). Why Melissa is so scary. *U.S. News and World Report.* Retrieved from www.usnews.com/usnews/issue/990412/12viru.htm

Morris, J. (2000). Net danger ahead? *PC Magazine, 19*(5), 74.

Network Ice. (undated:a). *FTP port probe*. Retrieved from advice.networkice.com/advice/Intrusions/2003004/default.htm

Network Ice. (undated:b). *SubSeven port probe*. Available from advice.networkice.com/advice/Intrusions/2003105/default.htm

Network Ice. (undated:c). *TCP port probe*. Available from advice.networkice.com/advice/Intrusions/2003102/default.htm

Network Ice. (undated:d). *TCP trojan horse probe*. Available from advice.networkice.com/advice/Intrusions/2003101/default.htm

Ricciardi, S. (1999). In pursuit of Internet intruders. *PC Magazine, 18*(21), 247, 252-253.

Rubenking, N. (2000). Freeware port scanners: Plug the holes. *PC Magazine, 19*(21), 208.

Scambray, J., McClure, S., & Kurtz, G. (2001). *Hacking exposed: Network security secrets & solutions* (2nd ed.). Berkeley: McGraw-Hill.

Schurman, K. (2000). When viruses attack. *PC Privacy, 8*(4). Available from www.smartcomputing.com/

Seltzer, L. (2000). Smarter security tools. *PC Magazine, 19*(6), 62.

Sengstack, J. (2000a, August 10). Insulate your PC from hackers. *CNN.com*. Available from www.cnn.com/2000/TECH/computing/08/10/diy.hacker.proofing.idg/

Sengstack, J. (2000b, September). Make your PC hacker-proof. *PC World*. Available at www.pcworld.com/howto/article/0,aid,17759,00.asp

Sheldon, L. (2000, August 28). Personal Firewalls. *ZDNet Reviews*. Available at www5.zdnet.com/products/stories/reviews/0%2C4161%2C2615071%2C00.html

Sonicwall, Inc. (2001). Untitled. *PC Magazine, 20*(4), 82.

Strauch, J. (2000, July 14). Personal firewalls keep intruders at bay. *PC World.com*. Retrieved from www.pcworld.com/resource/printable/article.asp?aid=17637

Sweet, M. (2001). Secure your browser & your e-mail client. *Smart Computing, 12*(?), 44-46.

Thomas, S. (1999, October 14). Home hackers. *U.S. News and World Report*. Retrieved from www.usnews.com/usnews/issue/991004/nycu/hackers.htm

Zetter, K. (2000, October 13). How it works: Viruses. *PC World*. Retrieved from www.pcworld.com/features/article.asp?aid=31002

Zone Labs, Inc. (2001a, February). Where's my identity? *The Zone*. Retrieved from www.zonelabs.com/CS/pastarticles/22001_whereidentity.html

Zone Labs, Inc. (2001b, October). Password protected. *The Zone*. Retrieved from www.zonelabs.com/CS/pastarticles/22001_passwordprotected.html

Section 3:

The CALL Classroom and Beyond

12

Network-Based Language Teaching in the Japanese Context

Douglas S. Jarrell
Nagoya Women's University

1. Computers and Japan

1.1 Japan's slowness to adopt computers and the Internet

There are a number of reasons why Japan has been slow to incorporate computers into education. Some researchers maintain that Internet use has grown slowly due to the expense of telephone connection charges. While this may be a contributory factor, there are other important cultural and technological issues to consider. Japan has traditionally placed much greater value on penmanship and calligraphy than western cultures do. At the same time, the Japanese typewriter never became the important office machine that it was in the West: it was an expensive, unwieldy machine with approximately 2,400 characters (Kida, 2001), and its steep learning curve brought meagre gains in speed and efficiency. Moreover, as the popularity of the personal computer (PC) grew in North America and Europe, a different machine, the word processor, came to dominate the Japanese market.

The appearance of the word processor at the end of the 1970s and its spread in the early 1980s (Kida, 2001) finally gave Japan machines that could replace handwriting. Instead of hunting among over 2,000 characters on a giant "typesetting" board, users could type a sentence phonetically, hit the conversion key and (generally) get the appropriate Japanese. The PCs of the time were considerably more expensive and the preinstalled Japanese word processing software was not as accurate, so those involved with putting out information saw no benefit in switching from word processor to computer.

While word processing rapidly replaced handwriting in business, its use was never considered necessary for education. During their first nine years of education, Japanese children were expected to learn to read and write over 1,800 characters. Traditionally, handwriting had been considered so important that children were sent to a special school to learn calligraphy. In the 1980s and 1990s students in secondary education and in many universities were still turning in handwritten papers.

A society that traditionally valued handwriting did not see the need for information technology (IT) in the classroom, and the many office workers and professionals, including teachers, who had mastered the word processor saw no immediate benefit in switching to computers. As a result, when the Internet went commercial, most Japanese had no hardware to connect to it. The Internet itself was not such an attraction to most Japanese consumers in any case because the bulk of information was in English.

1.2 Teacher attitudes towards CALL

The beginning of the 21st century finds the computer a standard feature at universities throughout Japan. Email is used for interfaculty communication, replacing memos and printed announcements of meetings, and it is rapidly becoming the major means of communication between individual researchers. Academic mailing lists allow researchers ongoing discussion in their special fields of research, and information is more accessible due to online academic databases and university library catalogues.

Nonetheless, in a 1999 small-scale study looking at perceptions of technological innovation in Japanese educational institutions, Lee (1999) records a number of negative perceptions of CALL. While the sample of 15 is too small to generalize from, it reveals attitudes which must be addressed if an innovation such as CALL is to be successfully introduced into the Japanese educational system:

1. Fear of computers was the predominant reason for refusing CALL.
2. Since teachers were unfamiliar with computers, they did not want to be embarrassed in front of students.
3. There is little empirically proven theory on CALL.
4. Computer literacy skills would take a lot of time to master for learners and there would not be enough time left for learning English. (Lee, 1999, p. 155)

This paper will discuss the first two points here and deal with the fourth in a later section on teaching computer skills. As for the third point, the growing body of research should overcome this objection in time.

The fear of operating a computer, like that of driving, disappears with time and effort. Computers and cars are both complicated machines that require practice to operate successfully. In the present educational environment in Japan, teachers have no choice but use computers so it is simply a matter of time before they lose their fear.

Embarrassment deriving from technical incompetence may be a stronger disincentive to use CALL, especially in an educational context such as that of Japan, where the teacher is seen more as the source of knowledge than as a facilitator of learning. If teachers appear to lack knowledge, they can lose student respect. This problem can be solved, however, if educational institutions provide support, ranging from computer-training workshops for teachers to computer-literate teaching assistants in the classroom. Besides, simple troubleshooting skills and an awareness of common problems (Mennim & Moore, 1998, pp. 41-42) are what teachers need to use a CALL laboratory.

1.3 CALL and linguistic diversity

Japanese university educators, confronted with an ever-widening range of student linguistic abilities, can benefit from certain advantages CALL has over the traditional classroom setup. Due to the continuing decline in the number of high school graduates in Japan, most universities are abandoning a strict selectivity in order to maintain enrolment. Recently formed *admission offices* accept students on the basis of an interview and a paper, both in Japanese, without a language test. As a result of the diversity of admission procedures, incoming students show extreme differences in linguistic ability – in the two-year college of Nagoya Women's University (NWU), IP-TOEIC scores of entering English majors in 2001 ranged from a low of 175 to a high of 440.

There is a long-held belief in Japanese education that fairness requires everyone to study the same courses regardless of ability, but, with a widening gap between the high and low level students within one class, this is no longer tenable. English departments around Japan have started to stream classes to satisfy the needs of such a diverse student body, but the number of such classes is limited due to scheduling conflicts and financial considerations. CALL may provide an alternative because of its unique one-on-one computer-student interface and potential to provide individualized instruction for each computer user.

1.4 Computer literacy among students

Until recently, incoming university students showed little familiarity with computers. An EFL teacher using computers was therefore responsible for teaching both English and computer skills. An experienced teacher of Internet-related EFL courses stated as recently as 1999 that, "for every dozen students, usually two or three have used computers before, and only one has a computer at home. Within those two groups, it is rare to have one student who has used the Internet" (Pellowe, 1999, pp. 203-204). This accords with my own experience in 2000: Few students had a computer at home, and only one or two had previously tried the Internet. The results of a questionnaire given to incoming English Department students at NWU in April 2001, however, showed a changing picture: all the students reported having used a computer previously although the number with Internet experience still remained low. A more comprehensive questionnaire administered in July 2001 to all the 110 students in the 2-year English Department showed that approximately 85% of second-year students and 70% of first-year students had computers at home, and most of these were connected to the Internet.

The students may have been unfamiliar with computers and computer-based email at the time of the questionnaire, but 100% reported owning cellphones, and 81% of the freshmen responded that they had previously used phone-based email. There was also an increase, compared to the previous year, in the number of students stating that they could type. One reason for this may be the growing number of students recruited from commercial high schools where the curriculum concentrates on office skills rather than academic subjects.

1.5 A growing need for computer skills

Some computer skills are now required for university and college students in search of jobs. According to career placement centre staff at NWU, major companies based in the Tokyo area began advertising job openings on the Internet in 1997. By 1999 companies throughout Japan had begun offering jobs online and were giving job seekers the choice of applying either by postcard (the traditional way) or by email. In the last year, however, many major companies have switched over entirely to email applications. The suddenness of this change reflects the speed at which the Internet is entering into everyday business in Japan.

The importance of the Internet to job-hunters is also reflected in second-year (graduating) student responses to the July questionnaire: 75% of students using computers outside the classroom were using them for job-hunting. The percentage of students using email, chat, or instant messenger was only marginally higher, 76%.

1.6 Cellphones, digital cameras, and computers

Japan is changing fast, and two of the technologies leading this change are the i-mode cellphone and the digital camera. The popularity of these two technologies coincides with a tremendous increase in consumer interest in the computer and the Internet. The cellphone, already a large and growing market by the mid-90s, boomed in 1999 when DoCoMo unveiled its newest product, i-mode, the equivalent of the Internet for cellphone. By April 2000 over 9 million people had subscribed to the service, and by 2001 there were 36.9 million subscribers (Ministry of Public Management, Home Affairs, Posts, and Telecommunications, 2001). The i-mode was so popular because of its familiar package, the cellphone, which included email. Another reason for its popularity is its Japanese-language Internet service, user-friendly for a population leery of the English-dominated Internet.

At the same time, rising sales of digital cameras gave the Japanese consumer another reason to invest in a computer. Digital cameras had become popular even though the picture quality was not equal to that of film photography and prints were considerably more expensive. Uploaded onto a computer, however, the same photos could be viewed, edited, and sent by email via the Internet without any extra cost.

The i-mode gave Japanese consumers a taste of Internet technology, and the digital cameras were an added incentive for people to get their own computer. The number of computer owners and people with Internet access is increasing rapidly, and Japanese is becoming one of the major languages on the Internet. These converging trends signal that Japan now is ready for computer-assisted language learning.

2. CALL in the Japanese context

2.1 Communicative approaches to ELT in Japanese high schools

It is necessary to examine existing educational practices in Japan first to determine how CALL can be used effectively in Japanese universities. The traditional junior and senior high school foreign language teaching used to emphasize reading and grammar, but in the 1995 curriculum drawn up by the Ministry of Education, there was an important change:

> Aural/Oral Communication A (OC-A), a new senior high school English core course started in 1995. The course is aimed at developing conversational English ability in the Japanese English as a Foreign Language (EFL) high school instructional setting, where word-to-word translation and grammatical explanation of written text have played a dominant role for over a century. (Miura, 2000, p. 6)

In reality, curriculum change does not always mean a change in teaching methods and materials. Citing upcoming university entrance exams, high school teachers in charge of third-year graduating students freely admit to ignoring curriculum objectives in order to concentrate on helping their students pass university entrance examinations.

Many oral/aural textbooks published for the Communication A, B, and C courses do not provide opportunities for students to engage in communication. Miura (2000) analysed the activity types and found that in 12 of 16 Ministry-approved textbooks, more than 50% of the activities were both noninteractive (i.e., students do not engage in a conversational exchange with one another) and noncreative (i.e., students are required to respond with predetermined answers). In the case of two textbooks, all activities were noninteractive and noncreative.

The textbooks analysed by Miura (2000) were published in 1995 so it is possible that more recent textbooks have added more interactive, creative activities. Nonetheless, at the start of the Ministry of Education's new curriculum, many course materials were ill-suited to the Ministry's declared aim of making high school English teaching more communicative.

2.2 Recent changes in university English departments
In the last decade, a shift has taken place in university English departments, although the impetus seems to come from market forces rather than from the Ministry of Education. Prospective students avoid literature-based English departments, choosing instead ones with an emphasis on communication, i.e., *Department of International Communication* and *Department of English Communication*. Students want to learn modern languages and expect practical results from their university education, namely the ability to use English in the real world.

2.3 Computers for communication: Tutors or tools?
For many years the computer in CALL has been seen as *tutor*, a source of exercises and an evaluator. The earliest computer programs "taught" with software "consisting of grammar and vocabulary tutorials, drill and practice programs, and language testing instruments" (Kern & Warschauer, 2000, p. 8). Much modern instructional software is still designed in this way although there is a greater emphasis on interactivity and attractive multimedia environments. The tutor leads students through the activities with its one-on-one student-computer interface allowing students to study at their own pace. This frees the teacher to move around and provide individual help.

While instructional software programs have a variety of activities and give users a large degree of choice in what they want to study, the materials have to be designed with "correct" and "incorrect" answers in mind. Japanese students have spent a large part of their secondary education passively absorbing correct English grammar and vocabulary. When they reach university level and study in order to use the language, they need to go beyond exercises and experience the features of real communication: unpredictability, negotiation of meaning, and hypothesis testing (Miura, 2000). The computer in its role as tutor cannot satisfy this need.

Kern and Warschauer (2000) discuss an alternative to the computer-as-tutor view, network-based language teaching (NBLT) which "represents a new and different side of CALL, where human-to-human communication is the focus" (p. 1). The computer is seen here in the role of tool rather than tutor:

> Computer networking allows a powerful extension of the computer-as-tool in that it now facilitates access to other people as well as to information and data. Computer networking in the language classroom stems from two important technological (and social) developments: (1) computer-mediated communication (CMC) and (2) globally linked hypertext. (Kern & Warschauer, 2000, p. 11)

NBLT, and in particular CMC, has several advantages. First, numerous researchers have remarked on the motivational strength of CMC (Warschauer, 1995; Dias, 1998; Freiermuth, 1998; Muehleisen, 1998). Secondly, CMC is relatively inexpensive, requiring no more than browser software. Finally, only low-level computer skills are needed for email and chat operations.

3. Design features for a NBLT for lower level learners

NBLT is not intended to be taught independently of other university language courses. CMC and hypertext provide students with a means to communicate with other students using language skills they have already developed. "CALL can be integrated into the curriculum as a medium that allows students to experiment with concepts taught elsewhere" (Stevens, 1992, p. 12).

3.1 Factors that affect teaching

When designing introductory materials for NBLT, it is important to take the following factors into account:
- Some students may be computer-illiterate and completely unfamiliar with the keyboard.
- Some students may suffer from a fear of computers.
- While a low student-to-teacher ratio is ideal, it is unrealistic in most universities. As a result, teachers will not be able to provide support for individual students and keep everyone in the class at task without clearly delineating tasks.
- All computer operation instructions are platform-dependent. Steps will differ depending on the particular combination of hardware and software available in the classroom.
- Lower level learners may be unable to cope with authentic webpages full of English text. According to Muehleisen, "Sometimes a reader has to plow through a screen full of unfamiliar words just to determine the topic of a

particular page. This is bad enough for people reading in their native language, but for students dealing with a foreign language, it can quickly become overwhelming" (1997, p. 38).

- Quick online connections are possible within the host country, but connecting all the students simultaneously to foreign interactive sites may be impossible. Delays that prevent some learners from completing the task or even accessing the requisite webpage must be avoided to maintain interest.

3.2 Guidelines
In view of the above factors, the following are design guidelines for NBLT:

- Class materials should teach basic computer skills in simple steps. Revision should be built into the course to allow students to familiarize themselves with computers and develop basic skills.
- Introductory typing skills should be part of the course. Poor typing will make any task more difficult and time-consuming.
- Initial lessons for a large class should be taught in lockstep fashion so that every student has the opportunity to learn basic computer and typing skills.
- Classroom support for students should include printed handouts, both as an initial guide for the lesson and for later reference. Teachers and their verbal instructions are often ignored because students are distracted by the visual interface with the computer.
- Bilingual vocabulary exercises are an efficient way to teach technical vocabulary. The completed exercises become a glossary of computer terms which the teacher can use in simple English instructions.
- Tasks should be designed for the particular combination of hardware and software available. Because of the constantly-changing nature of IT and equipment, CALL teachers need to keep abreast of improvements in equipment. Mistakes can result in the author losing face with students, a factor why novice Japanese teachers gave "be[ing] embarrassed in front of students" as one reason for their negative view of CALL (Lee, 1999, p. 155).
- If a Japanese operating system is used, specific problems must be addressed early on. First, students need to know when and how to switch the keyboard from English to Japanese and vice versa. Secondly, they need to understand the capability of browsers to "second-guess" URLs as they are typed. Otherwise students will be taken to a previously-viewed Japanese website such as <yahoo.co.jp> instead of to the English site <www.yahoo.com>.
- Simple webpages with a minimum of text should be prepared by the teacher for students to use while practising, accessing, and navigating. Students need to get used to computer and Internet concepts such as scrolling and hyperlinks on an uncluttered website before attempting to navigate an authentic one.
- Coming from a language that uses a different writing system, students can be overwhelmed by the sheer volume of written English on a site such as Yahoo. Webpages that are sparsely worded and visually clear should be used to orient novice users. ESL/EFL-specific sites such as *Activities for ESL Students* (Kelly & Kelly, 1995-2001) are often the most comprehensible. When students are given a webpage task, the teacher should err on the easy side. As students

progress, more authentic sites can be introduced and open-ended tasks assigned.
• Accessing foreign interactive URLs should be done sparingly to avoid freezes
  and slow-downs that can interrupt class momentum. Groupwork with up to
  three students can reduce the number of computers online and, hence, the
  danger of slow-downs.

3.3 Teaching basic skills

The immediate goal is not to teach instructional English but to familiarize students
with the operation of their computer as soon as possible. There is no need to maintain
an English-only policy if it prevents students from understanding computer operations, in which case L1 explanations are recommended.

Introducing typing software such as Mavis Beacon Teaches Typing (1999) is the
next step. The advantages of this typing software are: (a) its sparsely-worded visual
interface with clear operating instructions; (b) the variety of activities included: typing from text, dictation practice, and games; and (c) its low cost. Using typing software also familiarizes users with the layout of the keyboard and gives additional
practice in navigating through a variety of windows.

English-language typing software is preferable to the Japanese version because
users get additional input from the English instructions. In the long run, the linguistic
benefits of reading English windows using this software should outweigh the difficulties that arise from differences between English and Japanese keyboard layouts.

Making typing software a regular part of the class has other advantages not directly related to language learning. Students can work on mechanical skills in the
target language before they are expected to produce language. Another benefit is that
typing practice at the beginning of each class creates a buffer period during which
latecomers can enter without interrupting the class.

3.4 Introducing the Internet and email

Once students have mastered basic computer skills, they are ready to go online. They
should be guided to a simple website, incorporating the basics of hypertext such as
links, buttons, and scrolling bars. An example of a website that teaches Internet basics
clearly and simply is *Internet for International Communication* (Pellowe, 1997).

At the same time, students need to be introduced to email so that electronic communication becomes a regular means of communication between teacher and student.
Weekly email messages and assignments force students to access their accounts on a
regular basis, and questions from the teacher promote more individualized teacher-
student communication.

Student output will be a major factor in the teacher's final evaluation, so it is
imperative that students learn quickly how to use the email software. As Dahlin (1997,
p. 30) notes, "Assessment is built in to such an activity because, if the students follow
the steps correctly, the teacher should receive replies back from each of them." Students soon realize that computers do not allow for mistakes. They must type the address accurately in order for the email to arrive at the intended destination. In NWU,
students use email software that requires the writer to click two separate mailbox
icons to send an email. Although they are told the procedure during the explanation of
the software, someone inevitably forgets to click the second icon and cannot send
their messages. As Dahlin reminds us, "just because students are able to follow the

steps of a routine once does not mean they really know how to do it. The teacher should try to design activities that recycle and reinforce this information" (Dahlin, 1997, p. 30).

Emoticons, or *smileys*, part of the email genre, provide examples of cultural differences. The author sent everyone a message containing a story about himself using the smile :) and several other common English emoticons. Their assignment was to respond with a story of their own using emoticons. One student wrote back using Japanese emoticons common to i-mode users, confirming that cellphone email does influence some aspects of CMC. Not knowing that (*.*) meant "surprised" and (T_T) "sad," the author had to write back for clarification. This short cultural exchange anticipates the larger potential of keypal exchanges with people from other cultures.

3.5 Chat

Chat, unlike email, is synchronous CMC involving both input and interaction, two factors that are thought to facilitate acquisition of a language (Ellis, summarized in Kitao & Kitao, 1999, p. 18). Chat can be both a peer activity among members of the same class or an activity with complete strangers. In this paper we will examine the in-class peer activity.

One software option for chat developed in Japan is called LECS (Language educational chat system) designed by Harada and Yasuda at Kanto Gakuin University (1999). This software can be easily accessed over the Internet without any downloading. The teacher divides a class into groups of up to five students, and each student logs on to a chat session. The advantages of the LECS system include (a) software with bilingual directions, making it very easy for Japanese to use; (b) different coloured text for each speaker so that the threads of conversations can be easily identified and followed within each chat group; and (c) finished chats available for download for future reference by the teacher.

Online chat offers the participant certain advantages compared to face-to-face conversation. "Kern (1995) found that American students studying French attempted 2 to 3.5 more turns online than during normal class-wide discussion. He also noted that all students participated, not just the dominant ones" (Freiermuth, 1998, p. 83). Freiermuth notes problems between people of different cultures in face-to-face conversation, "When teaching intensive English to a group of international students, I heard repeated complaints by Japanese students that Brazilians were 'noisy,' making it extremely difficult for [the Japanese] to contribute . . . they lost numerous chances to contribute" (Freiermuth, 1998, p. 83). He points out that individual differences lead to unequal speaking time even among different learners in the same culture. "Even when a task had a very specific directive, one speaker generally controlled the conversation" (p. 82).

The results of chat sessions in the author's Internet English class confirm Kern's (1995) finding that there is general participation in chat discussions. Students were randomly divided into groups and used chat aliases to hide their identities. Each session was based on a discussion question chosen by the teacher in question form: "What kind of food do you like?" was the topic for Session One. "What does summer make you think of?" was the topic for Session Two. "What do you think of computers?" was the topic for Session Three.

In Session One, the four members of one group showed the least variation with 7,

7, 7, and 8 turns respectively. The greatest variation occurred in a group where one member contributed 21 turns compared to 10, 11, and 12 turns for the others. In Session Two, turns ranged from 6 to 12, 6 to 8, 7 to 10, 8 to 9, and 9 to 13 for five groups. In Session Three, the number of turns per group member had evened out, three groups showing a small range: 10, 10, 11, and 12 turns, 7, 7, 8, and 9 turns, and 9, 9, 10, and 11 turns. The greatest range was 9, 11, 11, and 14 turns.

Chat also encourages important conversation strategies such as addressing participants by name, volunteering relevant information, and asking questions. Students showed a willingness to use dictionaries and consult the teacher or the teaching assistant on English vocabulary and syntax, a marked contrast to their attitude in speaking classes. In their email some participants included clarification questions to partners about unknown words that the partners had used.

While students are using email, the teacher is free to circulate, observing, answering questions, and contributing suggestions and English equivalents for Japanese words. Individual attention to student problems is possible in other classes, but students are more reluctant to ask for help in the different environment of the speaking classroom. In conclusion, communication with peers via chat appears to be a highly motivating activity, one that makes CMC a particularly effective tool in EFL education.

3.6 Keypals

Warschauer points out, after an examination of online learning at four different colleges, that "although authentic communication is a necessary condition for purposeful electronic language use, it is not a sufficient condition . . . students later tired of communication tasks that they perceived as meaningless" (Warschauer, 2000, p. 56). Chat sessions within the classroom are a step towards authentic communication, but there is a danger of overuse. Shifting from in-class chat sessions to international chat sessions is not recommended, however, because (a) chat rooms for non-native speakers (NNSs) can easily degenerate into practice sessions for four-letter words, (b) many students do not yet have the fluency to handle synchronous communication with native speakers (NSs), and (c) much NS chat is a mixture of acronyms and slang, both difficult and inappropriate for most EFL learners.

Keypals can be a better option because the one-to-one relationship adds civility and responsibility to an exchange between two people. The asynchronous feature of email allows motivated students to maintain contact with NSs and the less-motivated or lower level students to "talk" with NNSs who are closer to their own level. Japanese students who communicate with foreign keypals can experience, often for the first time, the need to use English to communicate.

There are, however, several problems that arise with keypal communication. Keypals can be extremely fickle and stop responding, their interests may not match those of the student, and the difference in English level can make communication very difficult. One of the author's students, a diligent student with good English skills, found herself getting no response from her lone keypal while her neighbour continued to receive responses from four or five keypals every week. Others (all students were women) received email from pushy men who asked for photos and addresses. Many students find it difficult to respond, especially when the first email is long and difficult to read.

Several students had keypals from Japan, and one student maintained the longest exchanges with her Japanese keypal. Hers were also some of the longest messages written by any student, indicating that there are benefits to be reaped from exchanges by people from similar as well as from different cultural backgrounds

Although it is impossible to know why some keypals respond while others do not, Warschauer offers a number of suggestions that should lead to a more equal distribution of communicative opportunities:
- Get multiple partners;
- Do exchanges with several classes in different countries;
- Set up mailing lists so that students can write to everyone in the other class;
- Give assignments to interview keypals (Warschauer, 1995).

4. Conclusion

There were a number of reasons why Japan and its universities were slow to adopt CALL and the Internet, but computer technology is now an accessible, viable resource for university-level EFL teaching in Japan. The incoming student population is growing more heterogeneous in its linguistic abilities, and CALL may provide a partial answer for dealing with diversity: The student-computer interface allows greater student autonomy and hence more opportunities for individualized learning.

CMC in the form of chat and email is particularly well-suited to communicative language teaching. It allows students the opportunity for exchanges not possible within the traditional classroom, exchanges that are examples of true communication with unpredictable outcomes, requiring negotiation of meaning, engaging the students in hypothesis testing, and most important in a language classroom, requiring the use of the L2. CMC also has the advantage of encouraging lower level students to use the target language for communication in spite of their difficulties with language learning. In the end it is these students, sitting at their computers well after the bell rings to finish up an email to a keypal, who demonstrate the effectiveness of using networked computers for communication in the classroom.

References

Dahlin, E. (1997). Email in the classroom. In P. N. D. Lewis & T. Shiozawa (Eds.), *CALL: Basics and beyond. (Proceedings of the second annual JALT CALL N-SIG conference, Chubu University, Japan, May 31-June 1, 1997)* (pp. 27-31). Nagoya: Chubu Nihon Kyouiku Bunkakai.

Dias, J. (1998). The teacher as chameleon: Computer-mediated communication & role transformation. In P. N. D. Lewis (Ed.), *Teachers, learners, and computers: Exploring relationships in CALL* (pp. 17-26). Nagoya: Chubu Nihon Kyouiku Bunkakai.

Freiermuth, M. (1998). Small group online chat: The great equalizer. In P. N. D. Lewis, (Ed.), *Teachers, learners, and computers: Exploring relationships in CALL* (pp. 81-86). Nagoya: Chubu Nihon Kyouiku Bunkakai.

Harada, T., & Yasuda, T. (1999). Language educational chat system (LECS). Available from home.kanto-gakuin.ac.jp/~taoka/lecs/

Kelly, L., & Kelly C. (1995-2001). Activities for ESL students. *The Internet TESL Journal.* Retrieved from a4esl.org/

Kern, R., & Warschauer, M. (2000). Theory and practice of network-based language teaching. In M. Warschauer & R. Kern (Eds.), *Network-based language teaching: Concepts and practice* (pp. 1-19). Cambridge: CUP.

Kida, J. (Revised 2001). *Part five, wapro: Interface between the Japanese language and the computer.* Available from www.honco.net/japanese/05/

Kitao, K. S., & Kitao, K. (1999). Using chat to teach English. In P. N. D. Lewis (Ed.), *Calling Asia, The proceedings of the 4th annual JALT CALL SIG conference, Kyoto, Japan, May 1999* (pp. 17-20). Nagoya: Chubu Nihon Kyouiku Bunkakai.

Lee, S-I. (1999). The legitimacy of CALL in ELT. In P. N. D. Lewis (Ed.), *Calling Asia, The proceedings of the 4th annual JALT CALL SIG conference, Kyoto, Japan, May 1999* (pp. 153-157). Nagoya: Chubu Nihon Kyouiku Bunkakai.

Mavis Beacon Teaches Typing (version 10). [Computer software]. (1999). Novato, California: Broderbund.

Mennim, P., & Moore, P. (1998). The WWW as content in an undergraduate English Curriculum. In P. N. D. Lewis, (Ed.), *Teachers, learners, and computers: Exploring relationships in CALL* (pp. 35-42). Nagoya: Chubu Nihon Kyouiku Bunkakai.

Ministry of Public Management, Home Affairs, Posts and Telecommunications. (2001). Statistics. Available from www.yusei.go.jp/eng/Statistics/number_users2001july.html

Miura, T. (2000). A system for analysing conversation textbooks. *JALT Journal, 22*(1), 6-26.

Muehleisen, V. (1997). English via the Internet: Using the Internet in university classes. In P. N. D. Lewis & T. Shiozawa (Eds.), *CALL: Basics and beyond (Proceedings of the 2nd annual JALT CALL N-SIG conference, Chubu University, Japan, May 31-June 1, 1997)* (pp. 37-40). Nagoya: Chubu Nihon Kyouiku Bunkakai.

Muehleisen, V. (1998). Motivating language learning through email exchange. In P. N. D. Lewis (Ed.), *Teachers, learners, and computers: Exploring relationships in CALL* (pp. 69-74). Nagoya: Chubu Nihon Kyouiku Bunkakai.

Pellowe, B. (1997). *Internet for international communication.* Available from www2.gol.com/users/billp/course

Pellowe, B. (1999). Designing web pages to introduce EFL students to the Internet. In In P. N. D. Lewis (Ed.), *Calling Asia, The proceedings of the 4th annual JALT CALL SIG conference, Kyoto, Japan, May 1999* (pp. 203-206). Nagoya: Chubu Nihon Kyouiku Bunkakai.

Stevens, V. (1992). In M. C. Pennington & V. Stevens (Eds.), *Computers in applied linguistics: An international perspective.* Avon: Multilingual Matters.

Warschauer, M. (1995). *Email for English teaching.* Alexandria: TESOL.

Warschauer, M. (2000). Online learning in second language classrooms. In M. Warschauer & R. Kern (Eds.), *Network-based language teaching: Concepts and practice* (pp. 41-58). Cambridge: CUP.

13

Keypal Exchanges for Writing Fluency and Intercultural Understanding

Kazunori Nozawa
Ritsumeikan University

1. Introduction

Student field research projects are nothing new to the EFL/ESL classroom. For many years, teachers have been sending students outside the classroom to collect data for the completion of longer-term projects. Such projects are valued by students, who view their completion as evidence of learning, and by language teachers, who view them as opportunities for direct contact with language in real contexts.

The wide availability of computer-based language resources in Japanese universities has extended the capability of teachers to provide direct access to language in new contexts. Collaborative learning in CALL laboratories is becoming commonplace. Students engage in computer mediated (keypal) exchanges,[1] webpage development, web searches, online publishing, discussion lists, and collaborative projects with students from around the globe. This first activity, keypal exchanges, focusing on the year 2000 classes at Kyoto University, is of special interest in this study.

2. Brief review of literature

The following brief review of keypal literature is limited to only a few example studies from among the many undertaken throughout Japan over the past several years. Furmanovsky (1999a, 1999b, 2000) argues that an organized "Intercultural Keypal

Project" involving two English language teachers in two countries who match classes for an extended period of time is much more effective than two individual keypal projects. However, he notes that it is not easy to find a class of a similar size, with a similar academic calendar or similar goals, and students of a similar language proficiency level. Findings from keypal projects involving Japanese, Taiwanese, and Turkish EFL classes indicate that the expectation of reciprocity is an important motivational factor. Students expressed excitement at getting new letters from keypals or seeing their faces on a website, and disappointment when letters were not forthcoming. Furmanovsky (2000) also looks at one effort to locate, customize, and develop ideas and suggestions for classroom email projects that CALL teachers working in EFL environments have made available through their homepages. He describes the adaptive process by which a keypal for intercultural communication project was conceptualized and designed using materials downloaded from the homepages.

Dahlin (1997) and Muehleisen (1997, 1998a, 1998b) encourage teachers to get their own classes to try email exchanges and offer practical suggestions to help them do so. They argue that email exchanges can motivate language learners by giving them a chance to develop language skills while at the same time forming a relationship with a person from another culture.

Ueno (2000) investigates students' English use in email discourse and analyses the corpus. While they find that their corpus-based approach can be a useful way to examine student writings for certain characteristics, they report both quantitative and qualitative results, concluding that, without guidance, students tend to write about themselves, leading to superficial learning. Furthermore, they note that students often lack the specialized vocabulary necessary to describe Japanese traditions and practices.

Dickson (2000) applies intercultural communication theory to such aspects of computer-mediated communication (CMC) as paralanguage, de-individualism, and hyperpersonalisation in the context of a US-Japan virtual team that exchanged emails like a keypal exchange project. Her case study analyses more than 1,200 email messages between Japanese and American executive board members and suggests that Japanese participants tend to employ a more "aggressive" form of English while American participants soften their expressions during online discussion, and encourage open exchange of opinions. She suggests that EFL/ESL teachers should note that Japanese students from "high-context cultures" may have to adjust both culturally and linguistically to email, which is essentially a "low-context" medium.

Von Kolln (1998) provides a theoretical framework and describes practical applications of what she calls the *email tandem*, a structured interchange in a foreign language with a partner via the Internet. She views the interchange as a conceptual forum for the development of understanding about intercultural learning strategies, the appropriate use of the Internet, and learner autonomy.

Robb (1996), Kitao and Kitao (1996), and Cowan (2001) all provide valuable suggestions for finding keypal partners, understanding expectations, preparing students, and tracking progress, using examples from personal experience. They conclude that keypal exchanges can be extremely rewarding for students but note that success takes planning and should not be assumed, especially the first time one undertakes an exchange. For more information on similar projects, see Daniels (1998), Freiermuth (1997), and Ryan (2000).

3. Facility – CALL laboratory

Effective online learning requires up-to-date hardware (networked computers), software, and a physical environment conducive to learning.[2] This study was conducted in a lab with 56 networked desktop computers (Windows NT), one 40-inch television monitor, one OHC (overhead data projector), three SONY analogue and digital audio/video players, and a laser printer. The facility was in its final year of operation before replacement. However, it was sufficient for keypal exchange activities.

4. Description of keypal classes

Keypal exchanges were conducted in three sections (e323, e337, e346) of an undergraduate English language training course (English II). Students were in at least their second year and had different majors. Class lists did not stabilize until the fifth week of classes as students shuffled their course loads. As Table 1 shows, the final enrolment for the three English II classes was 114 and the average class size was 38. Such a large size made it difficult to provide adequate instructional support to individual students.

Table 1. Information of students in English II classes.

English II	Final No. of Ss	Attendance Rate (%)	Assignment Completion Rate (%)	No. of dropouts
e323	40	84.02	82.39	1
e337	41	88	82.59	4
e346	33	91.65	88.90	1
AVG	38	87.89	84.63	2
Total	114			6

Many email exchanges ran into difficulty after the first introductory messages. Students did not seem to have had much experience in written exchange and were consequently at a loss about what to write in the next letter. It became clear very quickly that keypal instructors should provide topics for email discussion by a specified date in printed or online syllabi. For each keypal class, both the text (*Dave Sperling's Internet Activity Workbook*, 1999) and online syllabus site <cw.prenhall.com/bookbind/pubbooks/sperling> were used to promote exchange.[3] In addition, one or two teaching assistants for each class were hired by the faculty to help students resolve technical problems.[4]

4.1 Annual schedule and assignments

A total of 27 instructional weeks were provided in the 2000-2001 academic year. 21 different topics were assigned for discussion and submission by email (See Appendix

A). The average completion rate of assignments as a whole was 84.63% (See Table 1), a satisfactory level considering the course requirement of 70%. Students who did not complete assignments satisfactorily were requested to redo and resubmit them.

The week 24 class was a MS-PowerPoint 97 workshop in which students learned basic operation of the program, including the use of templates. It was very useful for most students as well as the TAs who had less experience with the software.[5] Student online presentations followed in weeks 26 and 27 and were peer-evaluated.[6]

5. Result analysis and discussion

Although there were individual differences in information literacy among students at Kyoto University, generally speaking, their level was satisfactory and most completed the majority of required assignments. However, some students had difficulty finding a keypal even after several attempts through the website, and others lost partners after the first semester or could not get a response from them in time to complete an assignment. Both students and instructor became aware that these sorts of difficulties are quite general phenomena in keypal exchanges. In this particular case, the author ultimately had to replace a summary assignment with selected topics from the textbook.

5.1. Student reactions

In the final class hour of the second semester, students were asked to send in their comments on the keypal projects by email. Table 2 shows submission rates.

Table 2. Final comments on keypal projects.

English II	Final No. of Ss	No. of Ss submitting comments	Submitting Rate (%)
e323	40	28	70
e337	41	31	75.6
e346	33	26	78.7
AVG	38	28.33	74.
Total	114	85	

Typical positive comments from these students were as follows:
- The class was not like an ordinary English class and it was learner-centred so that I was motivated to complete the work and felt it interesting.
- The class seemed to be like the introduction of "information processing" that should be necessary for all the students in the 21st century.
- I learned the basics of computer and network system while completing the assignments.
- I could enjoy reading a variety of webpages in different fields and found the assigned activities very stimulating and fun.

- I strongly felt that my speed-reading and searching information skills seemed to be improved.
- The text content was a bit easy for me and quite understandable to complete the assignments.
- I was willing to attend the weekly class because I liked the Internet activities and assignments.
- Through completing the assignments, my reading and writing skills improved extensively.
- Learning about the presentation software (MS PowerPoint) was very useful and giving a presentation with it was a great experience.
- I enjoyed other students' presentations that were different topics and contents.
- Submitting the assignments via email was easy and the evaluation standard was clear and fair.
- I was so excited and felt so happy when I received a reply from a foreign student for the first time because I never had a chance to communicate with any foreign people.
- Through keypal exchanges I made some friends and enjoyed the communication with them while understanding their culture.

On the other hand, typical negative student comments were as follows:
- It was difficult to find an appropriate keypal who had the same objectives.
- Even after I found a keypal, I could not continue to exchange emails because the keypal simply did not reply my emails for various unknown reasons.
- I received some impolite emails from the students on the list of the website whom I had contacted and felt very uncomfortable with them.
- I could not find any keypals at all even after I tried to contact many listed students on the web so that I had to replace the summary reports with the textbook assignments
- It took much longer time to search and find an appropriate website and write a report every week than I thought.
- The preparation time for my presentation was not enough and I felt that it should be longer.
- There were too many assignments to complete and they should be less.
- I had a bit hard time to understand the teacher's instruction in English because my listening skill was still not high enough.

5.2 Questionnaire results

Below are the results of the questionnaire (See Appendix B for graphs) distributed to students at the end of the class for email response.

First (See Figure C1), 57 out of 68 respondents (83.82%) chose A (*very comprehensible*) or B (*fairly comprehensible*) for the written syllabus given at the beginning of the first semester.

Most students (73.53%) found using the textbook simultaneously with the website beneficial. If we interpret C as positive, almost all students were happy to use the textbook (See Fig. C2).

About half the respondents (52.94%) gave a clear positive evaluation for the 22 assignments. If the number of C responses (*neither interesting and fair nor uninter-*

esting and unfair number of assignments) is interpreted as positive, 53 out of 68 (77.94%) students were satisfied with the content and number of the assignments (See Fig. C3).

Moreover, as Figure C4 shows, individual keypal exchanges were not very successful. Only 22 out of 68 (32.35%) answered that they found more than one partner and made keypal exchanges regularly or properly. On the other hand, 28 students (41.18%) found one but experienced a sudden termination of the exchange. Unfortunately 18 students (26.47%) found it impossible to establish an exchange and had to replace the summary reports with exercises in the textbook. Some were very unhappy about this.

Only 10 out of 36 (27.78%) students who had regular exchanges answered that they would definitely or probably continue their keypal relationships. On the other hand, 23 out of 36 (63.89%) answered that they would not continue. (See Fig. C5) The reasons for this apathy could be cultural, technical, institutional, or a combination of all three. Clearly there is a need for further analysis of this phenomenon.

Student presentations were rated highly, with 59 students (86.75%) answering A (*very useful and practical*) or B (*fairly useful and practical*). If C (*neither useful and practical nor useless and impractical*) is interpreted positively, the majority of students (94.12%) were happy to make use of PowerPoint (See Fig. C6).

Evaluation standards appear to have been clear and fair enough for the students. As Figure C7 shows, 53 out of 68 (77.94%) answered A (*focused on the learner and very clear and fair*) or B (*fairly clear and appropriate*). On the other hand, only 4 (5.88%) and 1 (1.47%) out of 68 answered D (*a bit unclear and unfair*) and E (*very unclear and unfair*) respectively.

Finally, exchanges as a whole were satisfactory. Figure C8 indicates that a majority (39 out of 68, i.e., 52.94%) gave 8 points on a 1-10 point scale. The total average evaluation was 7.74 points.

6. Conclusion and recommendations

To sum up, most students found the written syllabus comprehensible, using the textbook simultaneously with the website beneficial, and gave a clear positive evaluation for the assignments. However, individual keypal exchanges were not very successful and some were very unhappy about this. Only few students who had regular exchanges answered that they would definitely or probably continue their keypal relationships, but the rest answered that they would not. The reasons for this apathy could be cultural, technical, institutional, or a combination of all three. Student presentations were rated highly and the students were happy to make use of MS PowerPoint. The evaluation standards appear to have been clear and fair enough for the students. The keypal exchanges as a whole were satisfactory.

If email as an Internet tool is considered a basic "tech" medium, it would still be very effective in a carefully planned CALL class. Keypal exchanges offer many opportunities for learning English language and intercultural exploration. Perhaps the most important benefit is the potential such exchanges offer for richer dialogues with the world community.

Each keypal exchange project should have good aspects of strategies for student orientation, syllabus design, textbook selection, and curriculum development.

Finally, to conduct more successful keypal exchange projects in EFL settings, the following three different types of projects would be helpful:

- Intra-class project: Introduction of email interaction activities within the class to be familiar with interaction while learning communication style and netiquette;
- Intra- or Inter-class project: Introduction of group- or class-oriented collaborative projects on topics to exchange opinions between classes rather than individual keypal projects;
- Inter-class project: Introduction of collaborative projects focusing on a joint product such as webpages with the same objectives.

This article has briefly described a way of keypal exchange in a networked PC laboratory, and has pointed out some findings. There are certainly many more sources of information available, both in print and online. *The Language Teacher*, the Japan Association for Language Teaching's monthly newsletter, published a series of seven articles on using the Internet in language classes, *Volume 20*(10) to *Volume 21*(6); these articles contain many useful suggestions and are available online (langue.hyper.chubu.ac.jp/jalt/pub/tlt/index.html/).

In addition to these specific works, readers are recommended to check online journals for EFL/ESL teachers, such as the *Internet TESL Journal* <iteslj.org> and *CALL-EJ Online* <www.clec.ritsumei.ac.jp/english/callejonline/index.html>.

Notes

1. Keypal means an email penpal. Some WWW sites provide the service of helping people find keypals who share their interests.
2. Unfortunately many computer labs in Japan are already deficient in at least one of these requirements and may deteriorate to the point of worthlessness if financial support for upgrading is not provided.
3. The website address was changed once after the first semester when some content, such as a huge number of registered student lists, had gone.
4. TAs should be ideally interviewed or tested before being hired, and be monitored for the classes, and finally assessed by the CALL teacher rather than all this being done by the faculty administration.
5. A few students with less computer literacy wished to have more time to learn advanced techniques while others worried about the difference of technological backgrounds. However, such things did not affect them much when creating a presentation file.
6. All the classes were divided into two groups for two presentation days and each group had 90 minutes, but it was better to have three presentation days to give students more transition time to prepare before making their actual presentations as well as evaluation time.

References

Cowan, M. Y. (2001). *Keypal dos and don'ts.* Retrieved from www.sabotenweb.com/ Bookmarks/about/mark.html

Dahlin, E. D. (1997). Email in the classroom. In P. N. D. Lewis & T. Shiozawa (Eds.), *CALL: Basics and beyond (The proceedings of the second annual CALL N-SIG conference, Chubu University, Aichi, Japan, May 31-June 1, 1997)* (pp. 27-31). Nagoya: Chubu Nihon Kyouiku Bunkakai.

Daniels, P. (1998). Designing a global classroom project. In P. Holden-Moses & Y. Iwata (Eds.), *Computers in foreign language education.* Tokai University Foreign Language, 104-113.

Dickson, C. C. (2000). Beyond netiquette: US-Japan virtual teams and intercultural communication. In P. N. D. Lewis (Ed.), *Calling Asia: The proceedings of the 4th annual JALT CALL SIG Conference, Kyoto, Japan, May 1999* (pp. 119-123). Nagoya: Chubu Nihon Kyouiku Bunkakai.

Freiermuth, M. (1997). Using the Internet to promote writing in an international English composition class. In P. N. D. Lewis and T. Shiozawa (Eds.), *CALL: Basics and beyond (The proceedings of the second annual CALL N-SIG conference, Chubu University, Aichi, Japan, May 31-June 1, 1997)* (pp. 89-95). Nagoya: Chubu Nihon Kyouiku Bunkakai.

Furmanovsky, M. (1999a). Intercultural communication through computer keypals, *Kokusai Bunka Journal*, Ryukoku University Kokusai Bunka Gakkai, 144-149.

Furmanovsky, M. (1999b). *Keypals for intercultural communication: Using course syllabi and other materials in teacher homepages to help design an intercultural e-mail exchange project.* Retrieved from www.world.ryukoku.ac.jp/~michael/docs/ keypals_article.html

Furmanovsky, M. (2000). Keypals for intercultural communication: Using teacher homepages to design an email exchange project. In P. N. D. Lewis (Ed.), *Calling Asia: The proceedings of the 4th annual JALT CALL SIG Conference, Kyoto, Japan, May 1999* (pp. 235-238). Nagoya: Chubu Nihon Kyouiku Bunkakai.

Kitao, K., & Kitao, S. K. (1997). *Keypal opportunities for students on WWW.* Retrieved from ilc2.doshisha.ac.jp/users/kkitao/online/www/keypal.htm

Muehleisen, V. (1997). English via the Internet: Using the Internet in university classes. In P. N. D. Lewis & T. Shiozawa (Eds.), *CALL: Basics and beyond (The proceedings of the 2nd annual CALL N-SIG conference, Chubu University, Aichi, Japan, May 31-June 1, 1997)* (pp. 37-40). Nagoya: Chubu Nihon Kyouiku Bunkakai.

Muehleisen, V. (1998a). Motivating language learning through email exchange. In P. N. D. Lewis (Ed.), *Teachers, learners, and computers: Exploring relationships in CALL* (pp. 69-74). Nagoya: Chubu Nihon Kyouiku Bunkakai.

Muehleisen, V. (1998b). *Organizing email penpal exchanges.* Retrieved from faculty.web.waseda.ac.jp/vicky/papers/email.html

Robb, T. N. (1996). E-mail keypals for language fluency. *Foreign language notes, Foreign Language Educators of New Jersey, 38*(3), 8-10. Retrieved from www.kyoto-su.ac.jp/~trobb/keypals.html

Ryan, K. (2000). *Recipes for wired teachers.* Nagoya: JALT CALL SIG.

Sperling, D. (1999). *Dave Sperling's Internet Activity workbook.* New Jersey: PHR.

Ueno, Y. (2000). Corpus-based analyses of email by Japanese college students. *JACET*

Bulletin, 32, 137-149.
von Kolln, G. M. (1998). Developing learner autonomy via the international e-mail tandem network. In P. N. D. Lewis (Ed.), *Teachers, learners, and computers: Exploring relationships in CALL* (pp. 75-80). Nagoya: Chubu Nihon Kyouiku Bunkakai.

Appendix A

Course outline

[1st Semester]
No. 1 (Week 1) Your self-introduction
No. 2 (Week 2) Your keypal's introduction
No. 3 (Week 3) Chapter 2 Activity 4 (Amusement Parks)
No. 4 (Week 4) Chapter 3 Activity 3 (Extinct Animals)
No. 5 (Week 5) Chapter 4 Combination of Writing Activity & Activity 4 (Architecture)
No. 6 (Week 6) Chapter 5 Activity 4 (Louvre Museum)
No. 7 (Week 7) Chapter 7 Activity 5 (City Visitor Info)
No. 8 (Week 8) Chapter 9 Activity 2 (Infamous American Criminals)
No. 9 (Week 9) Chapter 10 Activity 2 (Proverb Search), 3 (Folk Tale), & 4 (Gestures)
No. 10 (Week 10) Chapter 11 Activity 2 (Specialized Schools) & 4 (Top Universities)
No. 11 (Week 11) Chapter 12 Activity 2 (Family HP) & 3 (Wedding Customs)
No. 12 (Week 12) Chapter 13 Activity 2 (Favourite Food) & Activity 4 (Restaurant Review)

[2nd Semester]
No. 13 (Week 14) Chapter 14 Activity 2 (Names of Games) & Activity 3 (Computer/Internet Game)
No. 14 (Week 15) Chapter 15 Activity 2 (Explorers) & 3 (Top of the World)
No. 15 (Week 16) Chapter 16 Activity 3 (Ailments and their causes) & 4 (Exercise and Health Benefits)
No. 16 (Week 17) Chapter 17 Activity 3 (Chinese Dynasties) & Activity 4 (Historical Events in the United States)
No. 17 (Week 18) Chapter 18 Activity 2 (Holiday Search) & Activity 4 (Find that holiday)
No. 18 (Week 19) Chapter 19 Activity 2 (Author Search), Activity 3 (The Ten Best English Novels) & Activity 4 (Web-Published Stories)
No. 19 (Week 20) Chapter 20 Activity 2 (Foreign Currencies) & Activity 3 (Shopping Trip)
No. 20 (Week 21) Chapter 21 Activity 2 (The 10 Most Profitable Movies), Activity 3 (Upcoming Movies), & Activity 4 (Movie Star Search)
No. 21 (Week 22) Chapter 22 Activity 2 (World Music Search), Activity 3 (Billboard Top 10), & Activity 4 (Top Pop Oldies)
No. 22 (Week 23) Chapter 25 Activity 2 (Great Scientists), Activity 3 (Scientific Fields), & Activity 4 (Discoveries and Inventions)

Week 24 MS-PowerPoint Workshop
Week 25 Preparation for Individual Presentations including watching a video on Presentation Basics
Week 26 Individual Presentations & Evaluations (1)
Week 27 Individual Presentations & Evaluations (2)

Appendix B

The questionnaire (translated from Japanese into English)

Q1. Was the syllabus (objectives, schedule, assignments, and evaluation) given at the beginning of the first semester comprehensible?

Very comprehensible
Fairly comprehensible
Neither comprehensible nor incomprehensible
A bit incomprehensible
Very incomprehensible

Q2. What did you think of the textbook?

Good and colourful design and easy to understand
Appropriate design and ordinary content
Neither well-designed nor poorly designed
Poorly designed and uninteresting content
Very uninteresting design and content

Q3. What did you think of the quantity and quality of the assignments?

Very interesting content and appropriate number of assignments
Fairly interesting content and fair number of assignments
Neither interesting and fair nor uninteresting and unfair number of assignments
Somewhat uninteresting content and a bit too many assignments
Very uninteresting content and too many assignments

Q4. Have you found any keypals and exchanged emails?

I found more than one keypal and exchanged useful emails.
I found only one keypal and exchanged emails properly.
I found only one keypal but email exchanges were stopped suddenly.
I found no keypals and replaced the reports with the assigned exercises in the textbook.
I found no keypals but did not replace the reports with the assigned exercises in the textbook.

Q5. Will you continue to exchange emails with your keypal(s)? (Please answer this question if you answered A or B in Q4.)

Yes, we will definitely continue our keypal relationship.
Yes, we will probably continue our keypal relationship, but we are not sure.
Hard to say.
We will probably not continue our keypal relationship.
No, there is no possibility we will continue our keypal relationship.

Q6. What did you think of using MS-PowerPoint to create an online presentation?

Very useful and practical
Fairly useful and practical
Neither useful and practical nor useless and impractical
Useless and impractical
Very useless and impractical

Q7. What did you think of the evaluation standards?

Focused on the learner and very clear and fair
Fairly clear and appropriate
Neither clear and fair nor unclear and unfair
A bit unclear and unfair
Very unclear and unfair

Q8. As a whole, how would you grade this class on a scale from 1-10?

1 2 3 4 5 6 7 8 9 10

Q9. Further comments?

Appendix C

The results in graph form

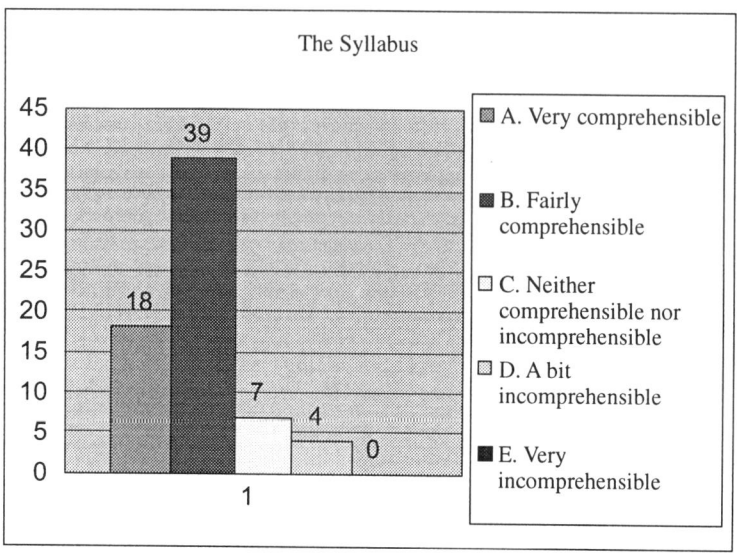

Figure C1. Syllabus comprehensibility.

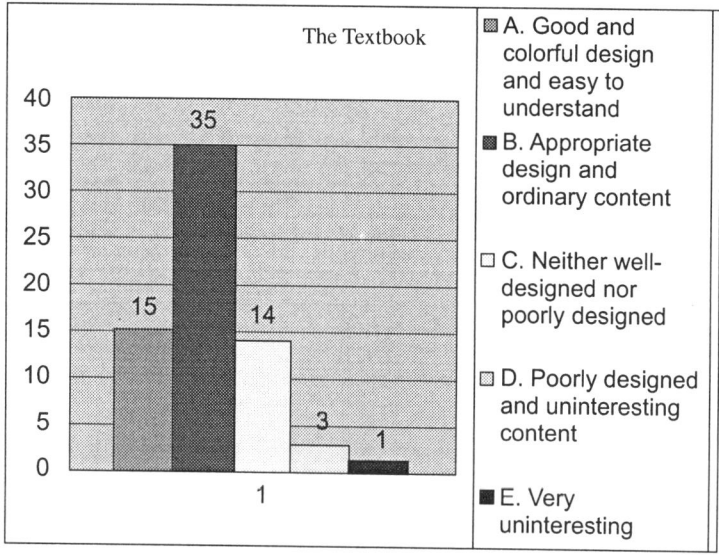

Figure C2. The appropriateness of the textbook.

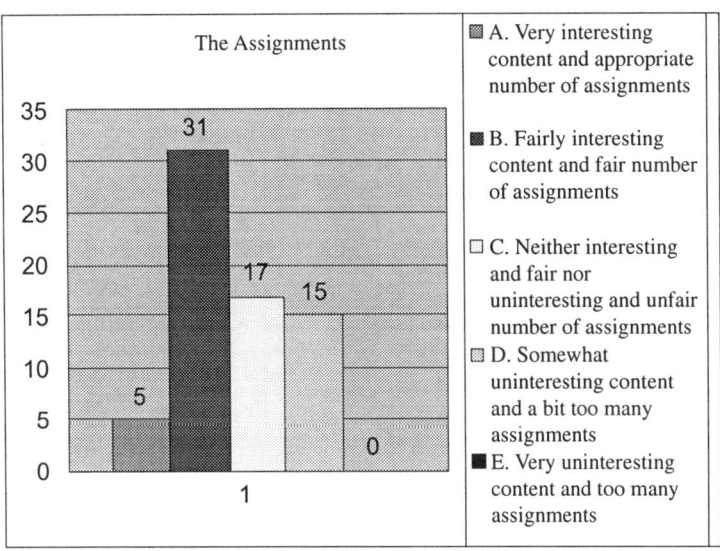

Figure C3. Quality and quantity of assignments.

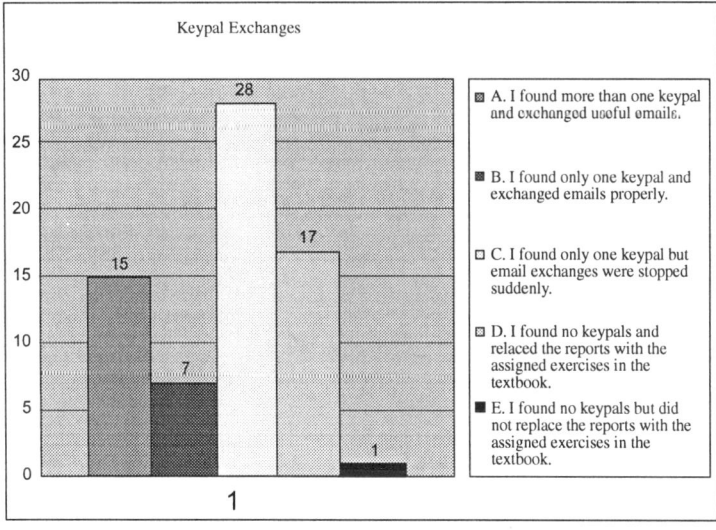

Figure C4. Keypal exchange results.

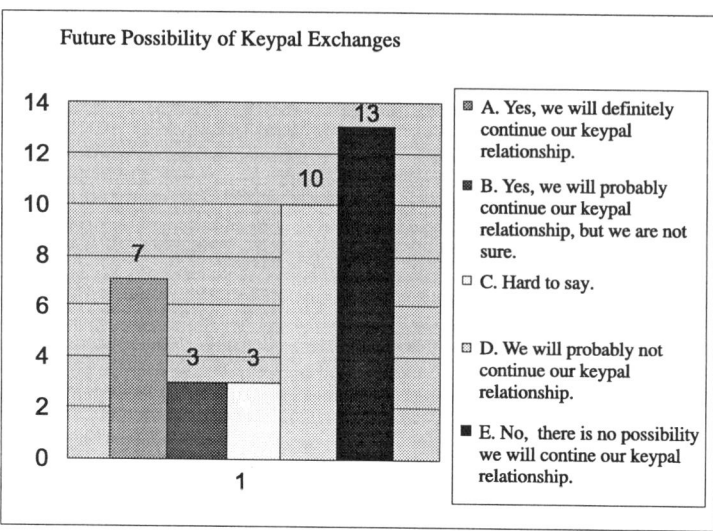

Figure C5. Keypal exchange possibility.

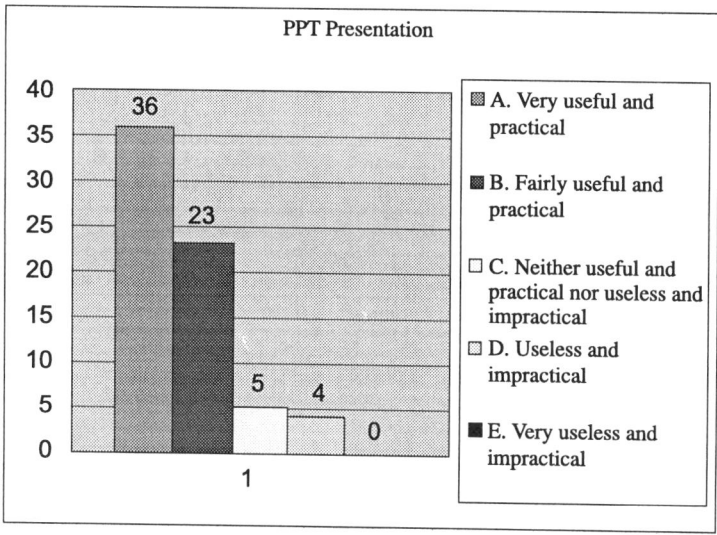

Figure C6. MS-PowerPoint and online presentation.

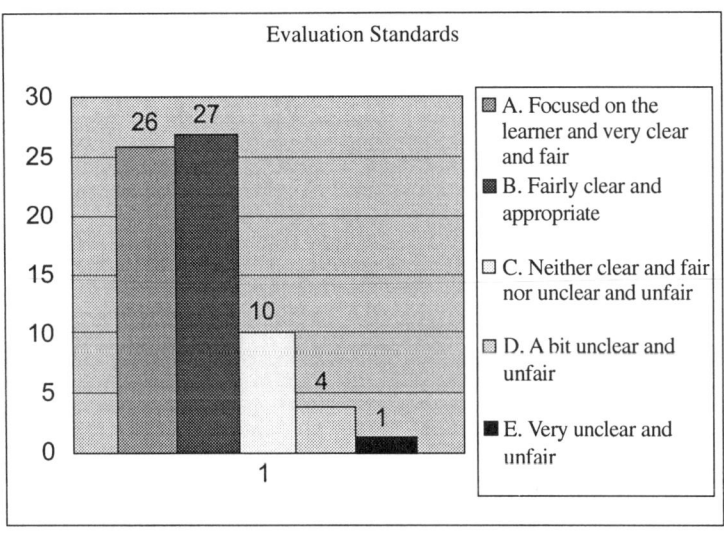

Figure C7. Evaluation standards.

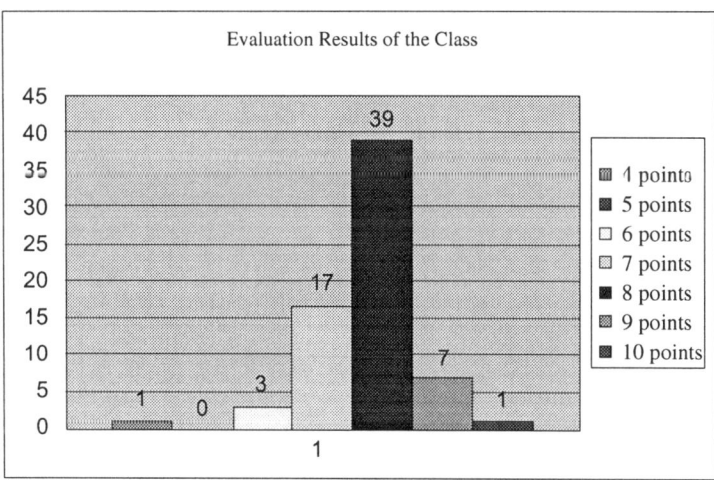

Figure C8. Evaluation results of the keypal exchange class.

14

Teaching Cultural Awareness Through Writing: Student Webpage Projects

Kenji Kitao
Doshisha University

1. Introduction

The Internet provides environments in which Japanese students can encounter authentic English and experience communication in English, which is difficult otherwise. Many colleges and universities in Japan have computer labs connected to the Internet already; all public primary and secondary schools should be connected by 2002, and the aim is to have a computer classroom with highspeed Internet connections in every school by the end of 2005.

Among the various learning activities utilizing the Internet, web projects are especially useful in motivating students to study English, teaching awareness of culture, and improving students' writing as well as their skill at researching and presenting information. Such assignments also help students develop skills, including formatting webpages and using search engines or links pages, as well as improve computer literacy, familiarizing them with software and Internet resources. These assignments involve choosing a research theme, finding and reading information, selecting useful material, organizing it, analysing the audience, and arranging everything in a clear, understandable fashion, all of which are basic to a university education.

This paper discusses the situation in Japanese schools regarding computer and Internet use, considers advantages and disadvantages of web projects, suggests ways to carry out these projects, and introduces some examples, in each case discussing final results. An appendix lists files useful for teachers wanting to try web projects.

2. Secondary schools and universities in Japan

To help place the projects in context, it is helpful to have some background information about secondary and tertiary English education in Japan. Secondary schools have 40 students per class. Public junior high schools have three hours of English a week, and senior high schools have four or five, though classes in private schools may meet more often. The school year averages 30-32 weeks.

Universities and colleges have two semesters; classes meet about 13 times a semester. Each class is 85 to 100 minutes long and most meet once a week. Classes range from 20 to 60 students. Most students take two English classes per semester for the first two years.

2.1 CALL at Japanese secondary schools and universities

Increasing numbers of secondary schools have computer labs with up to 40 Internet-linked computers. While there is heavy investment in equipment, almost no training is provided for English language teachers on using this technology in the classroom, and in most cases, no assistance with technical or pedagogical aspects either. Teachers familiar with the Internet, eager – or in some cases pressured – to do so, use computers and the Internet for English classes. However, even those comfortable with the technology may not know how to use it effectively in the classroom.

University facilities are better than those in secondary schools, and it is possible for many university students to use computers and the Internet fairly freely, including from home. However, even at the tertiary level, English language teachers experience a lack of training and help similar to that of secondary school teachers outlined above.

Some universities have started basic computer lessons for students, including netiquette and computer ethics. This makes it easier to use the CALL lab for English classes, since the need to teach computer skills is at least partly eliminated.

Since little training is available concerning CALL materials and use, most teachers are self-taught. Few universities in Japan have English teachers with CALL backgrounds, so most operate on a trial-and-error basis.

3. Using the Internet for instruction

The Internet features in English instruction in three ways. The first is via email, mailing lists, and chat or bulletin boards. Teachers need to demonstrate how to write messages and abide by netiquette conventions; Japanese students do not always follow directions, and may thus need supervision. Secondly, students use the Web to gather information. However, finding resources is often difficult and time-consuming, and needs high English proficiency to locate specific information and evaluate accuracy or potential bias. A resource link page may help. Thirdly, students make webpages – for example, short compositions with links, or pictures with explanations. Although making good webpages is difficult, Robb (2000) and Nozawa (1998) report on the effectiveness of their web projects, and a collection of about eighty websites made by English language teachers was recently published (Ryan, 2000).

3.1 Avoiding problems when using the Internet to teach English
There are a number of pitfalls for teachers to avoid when using the Internet for ESL. This section looks at some of these problems and how to avoid them.

The Internet itself should not be the centre of the classroom, but should be a tool to help students learn English. Teachers need a clear pedagogical purpose for using technology, and should limit its use to that purpose. It is unwise to use too many Internet functions in the same course, since they take time and energy to master and use efficiently in an English learning environment; English classes may become computer classes. Only Internet functions helping the goals of the English class should be used.

Teachers should provide specific and concrete directions, or even a manual on using the Internet, in easy English, for students to follow. Include copious examples, including some of what to avoid. Teachers need to practice using the Internet themselves to find the most efficient way for what they want students to do.

Initial use of an intranet can allow practice in a structured environment. For example, teachers can provide a limited number of webpages for student research. Students can do much without the Internet, including preparing for when it is available. Teachers must always be ready to cope with technical problems.

Many useful sites about the Internet in general and its application to ELT exist for teachers. A list of sites categorised in 20 groups can be found at *Site List* (See Appendix A for URL) (See also Kitao, 1998; 2001; Kitao & Kitao, 1997; 2001).

3.2 What are web projects?
There is a great variety of web projects and approaches toward them; one type, making English webpages, is effective for teaching language as well as performing research. In a typical assignment, students create a webpage and post it online for others to access. (See *Sample Projects*, Appendix A, for examples.)

3.3 Why use web projects in English classes?
Web projects are useful for English classes in Japan because:
- They make use of the ever more widespread computer facilities.
- They utilise limited class time efficiently. Once the assignment is explained and certain basic skills taught, most students work outside class, either individually or with a partner or group. Except for university English majors and senior high school students in special programs, English class time is very limited in Japan.
- Web project assignments can be used even by teachers unfamiliar with the Internet (Teeler, 2000, p. 77).
- Web projects can be used to teach writing, which is an underemphasised skill in English courses in Japan. In addition, these web projects can bring some of their benefits to traditional (i.e., non-CALL) writing classes.
- Students learn from a writer's point of view as they write, and a reader's point of view as they analyse their audience or read other projects.
- Projects are motivating. Students are interested in the Internet and computers and want to be able to use them. Projects related to different cultures can be especially motivating. Students find it encouraging to see their work on a website and get responses from readers, and this can lead to increased self-responsibility for their work.

- They allow students to encounter authentic reading materials, including vast amounts of English (Dunkley, 1997, p. 33).
- They lend themselves well to pairwork, groupwork, or peer criticism. More collaborative learning environments are fostered, reducing teacher work and helping improve instruction, especially in large classes.
- Students can learn research procedure, important in universities.
- Students can learn both oral and written presentation skills.

3.4 Types of web project

There are three basic types of web project: essay writing, making link pages for a subject, and a combination of both. Each type may include searching for information on the Net, reading and organising this information, writing up the final product, and illustrating it using photos or pictures.

For this type of project, three topics are possible: the students' culture, a target culture, or a comparison of both. The first is most common, but contains little information in English. The second one is good for research, but the audience for their final product will be Japanese, and writing English compositions for them is not "real" communication. The last is interesting but requires much work since it involves broad areas. Although all three have pros and cons, each type can be interesting and useful for English students; subjects related to cultures of various countries and enhancing international understanding are particularly appropriate (Johnson, 2001, p. 91).

Different projects require different levels of sophistication. The simplest type is a webpage containing short passages, some of which involve gathering information; the most sophisticated ones include research and extensive writing in English. Teachers should choose a level appropriate to time or student ability.

The typical web project includes, as a minimum, audience analysis, presentation, and writing. It is possible to include peer work such as asking questions, giving comments, critiquing, and making oral presentations about products. Students can revise their writing according to feedback from classmates and teacher; in effect, a process approach to writing.

The web project can include seeking information with search engines, evaluating webpages, scanning resources, selecting useful information, and organizing it in a way easy for readers to understand. It is a simple and practical way to learn about various cultures; it is often hard to search in traditional ways for relevant materials since Japanese university libraries have few such resources and secondary school libraries almost none.

Web projects allow Japanese students to search vast English resources, and offer them an opportunity to express themselves directly to English speakers, perhaps providing a rich environment for using English communicatively.

4. Advantages and disadvantages of web projects

4.1 Advantages

4.1.1 Communicating in English

Possibly the most important advantage of student-made webpages is that they enable students to use English as a form of communication. In EFL settings, students mainly use English in class for the teacher to evaluate (Robb, 1996, p. 7), generally reading no more than is assigned for a course. They often have no opportunity to read about subjects of choice or use English for communication. A webpage assignment can involve research both on and off the Internet, reading authentic materials in English, writing material for webpages, and reading and answering audience feedback; all involve communicating in English in meaningful ways.

4.1.2 Learning audience analysis
When making webpages that anyone might read, students need to be aware of their audience, and in particular, what that audience may already know or not know about the topic. For example, when making a webpage explaining Japan to non-Japanese, writers must consider how to get across aspects of the culture in English to a general reader with little or no knowledge of it.

4.1.3 Flexibility
Web projects are flexible and can be used with students of different levels, interests, and goals, as pair, group, or individual work (Gitsaki & Taylor, 2000, p. 242; Warschauer, Shetzer, & Meloni, 2000, p. 68). Low-level students can place a paragraph based on their knowledge or a picture description on a simple webpage. High-level students can do research and build a sophisticated website of several pages with links and photos, including sounds or animation.

Groups can cooperate on websites using self-chosen topics; this can be an advantage in large classes. Also, since Japanese college students meet only once a week, doing group-based web projects may improve collaboration skills (See Nozawa, 1997).

A web project can be long- or short-term, involving much time and energy or little. It could be the class main task over an entire semester or a minor project taking up one or two lessons; it is thus sufficiently flexible to fit in many types of English class in Japan.

4.1.4 Introducing students to web resources
As mentioned above, if the project involves research, a webpage assignment exposes students to available Internet resources, giving them an idea what is available, and showing them how to use search engines and links pages effectively.

4.1.5 Developing skills
The web project can include numerous types of language activity, thus developing a variety of helpful skills: using search engines, emailing, asking questions, giving comments, describing, analysing, illustrating, comparing, writing a research paper or essay, analysing an audience, avoiding plagiarism, or making effective presentations.

4.2 Disadvantages
Students may vary in the amount of experience they have had with computers, email, and browsers, and those with little experience will need much help. One possible solution is to form project groups with students of mixed computer skill levels. Solutions for a lack of familiarity with html are to use a template or webpage authoring program (See *Webpage Authoring Advice,* Appendix A, for suggestions). It is also possible to save a Word text file as html. The important point is for teachers to mini-

mise time spent producing html files, because this not the emphasis in an English class. The goal is for students to make a webpage that can be read comfortably.

Using search engines effectively is not easy. Students are likely to have difficulty finding useful web resources, and may need guidance.

Writing link descriptions/explanations or essays for English webpages may also be problematic. While unnecessary for teachers to completely correct the English, they should aim to choose topics within the students' range of expression.

Students may not yet have encountered a project where they choose a topic, gather resources, select and organize information, and present the final product effectively. They may not know how to critically read and evaluate the information they find. Again, teachers should be prepared to offer support where required.

5. Procedure for web projects

5.1 General procedures
Before beginning, it is important for teachers to consider the following points.
- What are the goals of the class? Does a web project fulfil those goals? What type of web project is most appropriate? What aspects of the web project should be emphasized?
- How much time is available to complete the project? If little class time is available, it should be reserved for giving instructions and, possibly, at the end of the project, for presentations; most work is done outside class. If more class time is available, student progress can be checked in class or peer assessment encouraged.
- How experienced and knowledgeable is the teacher in using computers and the Internet? For less experienced teachers, it is better to start with a simple project; experienced teachers might attempt more complex ones.

Before starting the project, if students are unfamiliar with the Internet, email, and the WWW, the teacher needs to introduce them briefly or problems will soon be encountered. Students need to know the basics of using a browser, a word processor, and email simultaneously, switching among them quickly, transferring from one to another, and copying URLs. (The following files are intended to help: *Internet, Using the World Wide Web, Email, Writing Materials,* and *Netiquette*; See Appendix A for URLs.)

Teachers need to explain the purpose, contents, and procedure of the web project, including how to choose a good topic, gather resources and read and critically evaluate them, choose relevant information (i.e., within the scope of the topic chosen), organize information, and write about the topic.

A good webpage topic should be concrete, somewhat familiar, and narrow enough that it can be covered in the length of the assignment. The following should help students (URLs for all the files can be found in Appendix A):
1. *Searching for Information on the Internet* explains the process of searching efficiently and finding and recognizing useful information. It also introduces a variety of search engines.
2. *Using AltaVista* gives specific advice for using this search engine.

3. *Evaluating Internet Resources* explains the importance of reading Internet resources critically and what factors students should investigate, including the information source, credibility, and possible bias.

Class procedure is important. When introducing the project, the teacher should give concrete directions and clarify expectations, show and discuss good examples of web projects, and outline the schedule for completion. Students need to cooperate, and also take into account both writer and reader viewpoints.

In the final lesson, students can present their projects, and practice making oral presentations while sharing work with their classmates. (*Public Speaking,* Appendix A, is intended to help teach oral presentations.)

Saving examples of web projects from previous years provides models for students. *Ikkyuji, Shimoda*, and *Doshisha* are examples of good projects and *Kyotanabe* shows a example of a poorly designed web project (Appendix A). Discuss criteria for good or bad results before students begin. Good projects are complete, concise, and accurate, with clear explanation of important points and well described pictures.

Though initial instruction is most significant, teacher feedback is important; not necessarily correction of English, but general suggestions and identification of problem areas. Web design and building is best set for homework, minimising class time.

In any type of project, teachers must emphasise that the final result will be on the Internet, accessible by people from all over the world. Thus, students are responsible for the final product, and their names should be affixed to it.

5.2 Essay type
The easiest project, suitable for even low level learners, is writing something short, concrete, interesting, and meaningful for both writer and reader. Guidelines for writing and an example are helpful. The teacher's goal is to database the students' writings efficiently. Even if each piece is short, a collection of writings on a related subject can be created.

5.2.1 Class practice: Dave Sperling's student project
Dave Sperling of California State University at Northridge had students interview each other and write brief biographies for a composition course (www.csun.edu/~hcesl004/CSUN.html). While this allows students to see their work on the Internet and can be motivating (Mak, 1997, p. 177), its usefulness for readers outside the class might be questioned. Nonetheless, the project can be used for peer work or class administration. Students write short self-introductions, information from an interview with a classmate, or on any assigned topic, and upload the results. Classmates read them, asking questions and constructively criticising, thereby communicating with the writer (See *CAI*, Appendix A, for details). The class as a whole can also discuss the reports, especially if posted on an intranet. (The files *Paragraph Evaluation* and *Project Evaluation*, Appendix A, offer help for essays.)

Even if the writing only consists of a paragraph or two, it is possible to make a database of many under the same broad topic.

5.2.2 Tom Robb's student projects
Tom Robb of Kyoto Sangyo University assigned his students a number of web projects

explaining Japanese culture (www.kyoto-su.ac.jp/~trobb/). These projects include *Famous Japanese Personages*, *Kyoto Restaurants*, and *Japanese Recipes* and *Kyoto Liquor Story* (both bilingual). The *Personages* database, the first and most extensive, contains brief categorised biographies of modern Japanese people, making a valuable resource. *Kyoto Restaurants* includes maps and illustrations. Student email addresses on many pages encourage reader feedback. While sophisticated web techniques make these databases more valuable for users, the English writing is not extensive.

5.2.3 Show and tell

Show and Tell is an easy project for any English class. Students provide photos and explain them in writing. Nakata's junior high school class project is available at <www.educa.nagoya-u.ac.jp/huzoku/n02.html>.

Ozeki Shuji at Nagoya Gakuin University includes *Studying Abroad*, *Japanese Traditional Culture*, and *Japanese Food*, on his students' webpage (www.intl.chubu.ac.jp/ozeki/ngu2000/miniBBS/). Anyone can submit a picture using an html form on the page.

5.2.4 Book descriptions

Book descriptions are especially appropriate for a reading course. Students report on graded readers, easy English books, or books about Japan written in English, including such information as the target audience, purpose, content, scope, and the student's opinion. A database of these can help future students choose books to read.

This project worked in an English class. Students view a list of books (e.g., *Books on Japan*, Appendix A) and discuss an example description; how it is written and its important points (for a sample, see *Handbook of International and Intercultural Communication*, Appendix A). Each student chooses a different book the first week, reads it, and writes a description over the next two weeks. After receiving peer- and teacher feedback, they revise it. (A page of links to sample descriptions is available at *Student Book Descriptions*, Appendix A.) This project used less than 30 minutes' class time.

5.2.5 Term papers

Some term papers are available online – students were warned at the outset that their papers might be read by many people. In some courses, we discussed the papers as students worked on them and in others, students just turned in the final draft. (See *Term Papers 1, 2, and 3*, Appendix A, for examples.) Writing longer essays might present more difficulties. Students may have problems selecting an appropriate topic, so the teacher could suggest concrete and familiar topics. Also, it might be helpful to provide some writing resources.

5.2.6 Doshisha project

After reviewing previous examples of student work, and assessing their good and bad points in class, students chose a topic related to Doshisha University. Example topics included: philosophy of education; the university departments; curriculum; Joseph Hardy Neesima (the founder); facilities; dining rooms; the campus; international programs; events; and club activities. Since students already knew something about the topic, little research was needed. Students read university literature, skimmed the webpage, or talked to staff. However, the main task was process writing.

After topic selection, students wrote a paragraph and then expanded it into several paragraphs. Each student read four peer essays, made comments, and asked questions to elicit desired additional information. Students then rewrote their essays, based on this feedback. The course concluded with an oral presentation. The entire process took six weeks, but required only three class periods in total. (The essays can be seen at *Doshisha*, Appendix A.)

Students found writing about the concrete and familiar easier than writing about the abstract and unfamiliar, and asking them to provide a photo may help make their writing easier to understand.

5.2.7 Japanese culture project

This more ambitious project involved a process writing approach over a semester (See *Culture 1, 2, and 3*, Appendix A.) First, students chose a topic on Japanese culture or Kyoto, the ancient city where Doshisha University is located. They then wrote a paragraph or introduction on that topic, followed by an outline, and created a first draft which they edited and revised. One specific task was assigned every week, following a strict overall schedule. During the project, students cooperated, gave feedback, and even corrected each other's errors.

5.2.8 Lilliam Hurst's student project

An unusual project was posted by students of Lilliam Hurst of College Claparede in Geneva, at: deil.lang.uiuc.edu/exchange/projects/geneva1.html. For a literature course, students read the play *A Streetcar Named Desire* and wrote letters to the characters in the play, offering sympathy for their situations or giving advice. Some students wrote letters the characters themselves might have composed. This project helps students understand the characters in a literary work, and offers readers interested in that work insights into students' points of view.

5.2.9 Tanaka Kazue's student projects

Students of Tanaka Kazue of Tokuyama Women's College completed a project on cross-cultural topics (www.tokujo.ac.jp/Tanaka/Kazue/kazue.html). They interviewed people in different cultures via email and researched Internet resources. While this project made sensible use of the Internet and allowed students to communicate with people of different cultures, unfortunately, the results, posted on student webpages, are only in Japanese. Having students write in English would make findings available to a wider audience and require students to work more in English.

5.2.10 Ruth Vilmi's student projects

Ruth Vilmi of Helsinki University of Technology uploaded student essays on various aspects of Finnish culture (www.ruthvilmi.net/hut/Project/Culture/). Some student web presentations also involved surveying other classes. This interesting resource for anyone interested in Finnish culture might be more useful if the topics were categorized.

5.2.11 Oguri Seiko's student projects

Oguri Seiko of Chubu University organised a student project, *Writing about Japan*, where students write about different aspects of Japanese culture in the categories of school, home, society, traditions, food, and seasons (www-clc.hyper.chubu.ac.jp/oguri/

japan/write_jpn.html). In addition, Oguri posted another of student writings about intercultural experiences in both English-speaking and Asian countries (langue.hyper.chubu.ac.jp/seiko/xculture94.html). Some essays are general, and others deal with specific issues such as inconveniences or surprises. The former is a potential resource for foreigners wishing to learn about Japan, or Japanese interested in how their culture might be described in English; the latter is useful for its descriptions of cultural differences from the viewpoint of Japanese students.

5.3 Link type

When making a links page, students post one topic with a number of links to subtopics. They gather 5-20 links for each subtopic using search engines, other links pages, and Net surfing. Adding a brief description and comments to each link involves much searching and reading of resources, so some Internet skills are necessary. Students enjoy this project type, and can make useful pages while honing web searching skills.

This web project involves searching for information, evaluating web resources, reading vast amounts of information in English, selecting useful and valuable webpages, and organizing links in subtopics. If students briefly describe each link in English, the assignment can also include some writing.

5.4 Combination type

This is the type the author has used most frequently with students. It involves making a links page with an introduction and conclusion or description. An optional explanation of this links page might include its purpose, target audience, contents, organization, or suggested use. This is a highly effective instruction method for developing Internet, research, and writing ability in English because it combines so many skills. (See *Projects, Study Abroad, Various Cultures and Visiting Countries,* and *Intercultural Communication,* Appendix A, for more information.)

One lesson was spent explaining the assignment procedure and viewing examples, and half a class was required to check each student's progress. The final class featured presentations.

Doing this project was a positive learning experience for students. Due to the opportunity for peer and teacher feedback, students were able to improve the quality of their writing progressively. Progress could be charted by comparing their original product with the finished one. Students took the project seriously, because they realized that the final product would be placed permanently on the Internet and accessible by people outside the class. They took into account the reader's viewpoint and were motivated to explain their ideas more thoroughly.

Students had difficulty using the Internet effectively and writing well at first, but by the end of the course, they had developed those skills and felt satisfied by their progress. They enjoyed working with other students, and this promoted cohesiveness in the classroom. Overall, students were happy that the skills learned in this class would be useful in their future.

5.4.1 Study abroad

The goal of this project was to make webpages on a certain university for students interested in studying there. Students each chose one university of interest, then selected ten webpages containing helpful information about the university: academic information, accommodation, student activities, assistance for international students, or English language requirements. Students made a quick fact page with enrolment, number of faculty, ratio of faculty to students, tuition, campus size, etc. Students carefully read, evaluated, and selected useful webpages about the university's location (town, state, country, for example), and transportation.

Students next described each site briefly in English. Any Japanese webpage links were placed on a separate page. Finally, they added an introduction, ending up with a page containing approximately 50 links with brief descriptions over five or six sections, in addition to the introduction and quick fact page. Procedure for other similar projects, such as *Various Cultures and Visiting Countries* was identical.

6. Conclusion

With the increasing availability of computers and Internet connections in secondary schools and universities in Japan, it is important to find uses for this technology that meet the needs of English language students and the goals of English courses. Web projects is one such use. They can be adapted to many different levels of English proficiency, topic areas, and course goals; therefore, they can be a powerful and effective way to teach English, even for those unfamiliar or underconfident with computers and the Internet.

Note

This work was funded by Doshisha University's Research Promotion Fund, 2000-2001, and a Grant-in-Aid for Exploratory Research, 1999-2001, from the Japan Society for the Promotion of Science.

References

Dunkley, D. (1997). Internet-derived material in the classroom. In P. N. D. Lewis (Ed.), *CALL: Basics and beyond (Proceedings of the second annual JALT CALL N-SIG conference)* (pp. 33-36). Nagoya: Chubu Nihon Kyouiku Bunkakai.

Gitsaki, C., & Taylor, R. P. (2000). Helping students create an English homepage. In P. N. D. Lewis (Ed.), *Calling Asia: Proceedings of the 4th annual JALT CALL SIG conference* (pp. 239-246). Nagoya: Chubu Nihon Kyouiku Bunkakai.

Johnson, T. H. (2001). World Wide Web: Cultural understanding at your fingertips. In Gaikokugo Kyoiku Media Gakkai (Eds.), *The annual conference of Japan Association for Language Education and Technology; Toward a harmonious multilingual and multicultural society in the 21st century; Collected papers* (pp. 91-93). Gaikokugo Kyoiku Media Gakkai.

Kitao, K. (1998). *Internet resources: ELT, linguistics, and communication.* Tokyo: Eichosha.

Kitao, K. (2001). Internet osusume site [Internet sites which I recommend]. *Eigo Kyoiku* [The English Teachers' Magazine], *50*(4), 17-19.

Kitao, K., & Kitao, S. K. (1997). *Eigo kyoiku no tameno pasokon to Internet: Yori kokat, eki na eigo kyoiku o motomete* [Personal computers and the Internet for English language teaching: Seeking more effective English language teaching]. Tokyo: Yohan Shuppan.

Kitao, K., & Kitao, S. K. (2001). *On-line resources and journals: ELT, linguistics, and communication.* Available from ilc2.doshisha.ac.jp/users/kkitao/online/

Mak, S. (1997). A multimedia autobiography in 60 minutes. In T. Boswood (Ed.), *New ways of using computers in language teaching* (pp. 177-179). Alexandria, VA: TESOL.

Nozawa, K. (1997). WWW project ni yoru kyodo gakushu to jibunka rikai [A WWW project using collaborative learning to understand your own culture]. In P. N. D. Lewis & T. Shiozawa (Eds.), *CALL: Basics and beyond (Proceedings of the second annual JALT CALL N-SIG conference)* (pp. 97-104). Nagoya: Chubu Nihon Kyouiku Bunkakai.

Nozawa, K. (1998). The world wide web projects through collaborative learning. *Ritsumeikan Keizaigaku* [Ritsumeikan Economics], 47(2, 3, & 4), 526-534.

Robb, T. (1996). The web as a tool for language teaching. *Kansai Chapter, The Language Laboratory Association of Japan, Monograph Series, 6,* 1-11.

Robb, T. (2000). Teaching writing with web projects: Famous personages in Japan. In E. Hanson-Smith (Ed.), *Technology-enhanced learning environments.* Alexandria, VA: TESOL.

Ryan, K. (2000). *Recipes for wired teachers: Practical ideas by teachers for teachers organized through the Japan Association for Language Teaching Computer Assisted Language Learning Special Interest Group.* Nagoya: Chubu Nihon Kyouiku Bunkakai.

Teeler, D. (2000). *How to use the Internet in ELT.* Essex: Longman.

Warschauer, M., Shetzer, H., & Meloni, C. (2000). *Internet for English teaching.* Alexandria, VA: TESOL.

Appendix A

Support page

The author has made a support page for this article which includes links to various useful sites for teachers interested in doing web projects as well as all URLs referred to in this paper. This support page can be found at: ilc2.doshisha.ac.jp/users/kkitao/library/article/call/web-project2.htm. If your browser is not capable of reading Japanese characters, use Shodouka (web.shodouka.com/) to read them.

URLs for the files listed in the article

All the URLs listed here need to be prefixed with the following server and directory address: ilc2.doshisha.ac.jp/users/kkitao/

Thus, the file *Site list*, for example, can be found at the full URL:

ilc2.doshisha.ac.jp/users/kkitao/japanese/library/article/eigokyoiku/internet.htm

Site list: japanese/library/article/eigokyoiku/internet.htm
Sample projects: japanese/online/project.htm
Webpage authoring advice: online/www/kitao/int-www.htm#adv
Internet: online/internet/art-inte.htm
Using the World Wide Web: library/article/call/www.htm
Email: library/article/call/email.htm
Writing materials: library/student/writing/
Netiquette: online/internet/art-netiquette.htm
Searching for information on the Internet: class/project/searching.htm
Using AltaVista: class/project/AltaVista.htm
Evaluating Internet resources: class/project/evaluating.htm
Public Speaking: library/student/public.htm
Ikkyuji: japan/ikkyuji/
Shimoda: et/shimoda/s3.htm
Doshisha: doshisha/
Kyotanabe: japan/kyotanabe/
CAI: class/cai/
Paragraph evaluation: class/project/paragraph.htm
Project evaluation: class/project/project.htm
Books on Japan: japanese/library/resource/book/japan.htm
Handbook of international and intercultural communication: library/resource/inter-
cultural/masante.htm
Student book descriptions: japanese/library/resource/book/project.htm
Term papers 1: class/meta/1999s/
Term papers 2: class/project/2001/meta/comm.htm
Term papers 3: class/meta/
Doshisha: class/project/2000/doshisha.htm
Culture 1: class/project/2000/japan.htm
Culture 2: class/d410/
Culture 3: class/essay/
Projects: class/material/project/
Study abroad: class/practicum/
Various cultures and visiting countries: class/meta/f/ and class/meta00/
Intercultural communication: class/special/

15

The Computer Based TOEFL and the Advent of Computerised Public English Testing in Japan

Michael Kruse
Chukyu University

1. Introduction

One of the most popular public examinations for learners of English in Japan has been the Test of English as a Foreign Language (TOEFL), produced and administered by Educational Testing Service (ETS), a commercial organisation based in Princeton, USA. Japan has had the largest TOEFL volume of any country in the world; for example, in 1996-97, Japanese TOEFL candidates accounted for just over 16% of the international total (Douglas, cited in Kruse & Shaver, 1998, p. 19). The TOEFL is designed to test whether candidates' grasp of English is sufficient for them to follow courses of study at universities in the United States. The normal pattern in most countries is for candidates to take the test only when ready to leave for study abroad, by which time most are graduates. However, Japanese candidates take the TOEFL for a variety of reasons. Some companies in Japan recognise a TOEFL score as evidence that a prospective employee is able to function in an English business environment; some universities, including respected public and private ones (such as Nagoya and Waseda, respectively), give academic credit for TOEFL scores. Finally, the majority of candidates in Japan are undergraduates, who take the TOEFL "for practice and just to see how they'll do" (p. 19). Correspondingly, the average Japanese TOEFL score has typically been 30 points lower than the international average.

It was originally planned that a new Computer Based TOEFL (CBT) would completely replace the familiar pencil-and-paper TOEFL, worldwide, by July 2000. In fact, ETS, creator and proprietor of the CBT, and Sylvan Learning Systems (Sylvan), its international agent for administration and marketing of the CBT, postponed introduction of the test in some countries. Not until spring 2001 did the CBT become the only internationally recognised form of the TOEFL available to candidates in Japan. Perhaps the large and highly profitable volume of less able candidates in Japan gave ETS and Sylvan pause. In any case, educators in Japan have been thankful for the breathing space in which to prepare candidates for the first public EFL examination in Japan administered by computer, especially as it seems significantly more difficult than the former paper-and-pencil version.

2. Computerised testing versus computer-adaptive testing

In considering the impact of the CBT on EFL education in Japan, it is important to distinguish mere computerisation from computer-adaptive testing (CAT). *Computerisation* is taken here to be a general term that refers to the act of performing on computer what could equally well be done without one; for example, simply writing a letter. *Computer-adaptive testing* is a technical term with a highly specific meaning. According to Dunkel,

> Second language (L2) computer-adaptive testing (CAT) is a technologically advanced method of assessment in which the computer selects and presents test items to examinees according to the estimated level of the examinee's language ability. (1999, p. 1)

CAT is an application of the principle that an examiner will make the most accurate assessment of a candidate's abilities by testing the candidate as precisely as possible at the candidate's *current level* of ability (Wainer, 1990, p. 10). In other words, if one wished to assess an interlocutor's level of English, one might begin by asking them a question one would expect them to be able to understand; one would then ask progressively harder or easier questions according to how capably the interlocutor replied. If the questions became too difficult, one would adjust them in the direction of simplicity, and vice-versa, thus gradually arriving at a highly refined assessment of the interlocutor's level. CAT attempts to simulate this process by computer.

Clearly, CAT will depend upon the availability of a specific type of software. Indeed, the Listening and Structure sections of the CBT incorporate a proprietary algorithm that assesses a candidate's performance on given questions, and assigns fresh questions considered to be of challenging but manageable difficulty according to its assessment of candidate performance thus far. Each question is preassigned to a certain difficulty level by ETS, and all candidates are presented at the outset with questions judged of moderate difficulty. A candidate's final score in these computer-adaptive parts of the test is ultimately determined by three factors: the number of correct responses, the level of difficulty associated with questions correctly answered, and candidate response times. Thus, if two candidates correctly answered the same questions, but one were slower than the other, the faster candidate would receive a higher score. On the other hand, if two candidates correctly answered the same *number* of questions in the same time, but one had correctly answered more difficult questions, the more knowledgeable candidate would score more highly.

It would be interesting to be able to discuss here the particular CAT software used in the CBT, together with the criteria by which ETS assigns levels of difficulty. Unfortunately, the exact nature of the algorithm involved remains a trade secret (from the point of view of ETS and Sylvan, the CBT is, after all, a commercial venture). However, its design *may* have been influenced by heuristics demonstrated and discussed by Stocking and Swanson in a number of publications in the 1990s (1993, 1998; and Swanson & Stocking, 1993), to which the speculatively inclined reader is referred. Their weighted deviations algorithm, for example, performs a function similar *in principle* to the CAT parts of the CBT, and their work is broadly acknowledged to have been highly influential in the field. In any case, it should be noted that research and development of CAT in general was first undertaken as long ago as the 1960s by the U.S. Department of Defense, keen to exploit CAT to assess a range of human abilities in military contexts. Thus, CAT is not a very recent innovation, and indeed there have already been a number of successful attempts to apply it in the field of EFL. For a review of CAT in a variety of educational fields, see Carlson (1994). From the point of view of EFL CAT software development, the CBT seems a successful but fairly modest achievement by current industry standards.

3. The effect of CAT in the CBT

There can be little doubt that incorporating CAT into the TOEFL has made life easier for candidates, at least in that it has made the test seem less threatening. One benefit of the computer-adaptive format is that candidates need answer fewer questions to receive an accurate assessment of their abilities, so the structure section of the CBT is considerably shorter than in the former TOEFL. Another benefit is that the CBT (minus a written component, which must still be marked separately by human examiners) can be scored automatically during administration of the test, and candidates receive a close approximation of their ultimate score immediately upon completion of the last question. There need no longer be a nail biting few months' wait for results. Finally, candidates receive questions in the CAT parts of the test in a sequence progressively tuned to their individual abilities: there are no shocking jumps from easy to difficult. Even weaker candidates are likely to receive strings of questions that they are able to answer with confidence. This may have the unintended side effect that a candidate is occasionally disappointed to receive a score lower than anticipated from their performance during the test, but the consensus seems to be that CAT will make taking the TOEFL less of an ordeal. According to Madsen (1986, p. 41), candidates found taking computer-adaptive ESL placement tests "an overwhelmingly positive" experience in comparison with taking written tests. Morrison quotes an actual CBT candidate:

> I took my test in a quiet room with just a few other people. We each had our own space with computer, chair, light, and headphones. It was really comfortable and easier on my nerves. (1999, p. 31)

4. Effects of computerisation

The candidate's comment above is doubly noteworthy as it suggests a perspective from which the incorporation of computers into the TOEFL is typical of that of CALL in general. That is, many significant consequences of newly computerised formats, certainly from a language learner's point of view, have less to do with the more dramatic potentials specific to computers, and more to do with the enhancement of mundane practicality. In other words, many developments apparently due to the advent of CALL do not represent new ideas so much as the application of old ideas in computerised contexts. For example, the features distinguishing the CBT from the former TOEFL of most significance to candidates – and this paper concerns chiefly Japanese candidates – are simple accompaniments to computerisation, not the result of any subtle exploitation of software or hardware.

The CBT is taken in an individual booth, at a time more or less of the candidate's own choosing. At least to some extent, volume, brightness, and the speed at which questions are given and must be answered, are brought under the candidate's control. These simple practical enhancements, together with the incorporation of CAT, reflect "the new test's overall approach to be more personalized" (Morrison, 1999, p. 31). In addition, the test material environment has become satisfyingly richer. It now includes images and graphics, both as simple accompaniments to dialogues, and as actual question or answer cues. These may all appear welcome changes, but for Japanese candidates, there is reason to believe that some may be less so. The CBT affords numerous opportunities for the prevaricator to come to grief. Candidates have more control of the test environment, and those of a naturally hesitant character may exercise too much caution in seeking to control the speed, although, as we have seen, there is a trade-off between answer scores and time. Japanese students are not noted for haste or decisiveness, and computerisation of the TOEFL has resulted in candidates having to make more decisions for themselves, some of which (such as volume and speed) are largely administrative in nature.

Even if the CBT were simply the familiar paper-and-pencil test transferred to a computer, this alone would pose additional problems for Japanese candidates. The introduction of computers into schools in Japan has occurred at a rate much slower than in the West (Pelgrum & Plomp, 1993, p. 323). Indeed, a member of the team who developed the CBT observed, even while the CBT was still in development, that, according to Douglas:

> We are all aware that computer use in Japanese schools lags behind that of most other developed nations . . . So, Japanese students are going to be at some disadvantage when it comes to the computer-based TOEFL. (Kruse & Shaver, 1998, p. 19)

In fact, computer use does not correlate so neatly with conventional economic measures. India, for example, has established a reputation for computer sophistication well in advance of its general economic development. Nonetheless, it is safe to assume that many candidates for the TOEFL do come from nations which have a computer literacy rate even lower than that of Japan. ETS has attempted to redress the imbalance in candidates' familiarity with computers by incorporating a half-hour tutorial in basic computer skills as a preliminary to the test itself, as well as a brief tutorial at the beginning of each section of the test. These tutorials use simple language and graphics to present the computer skills needed to answer questions, and then demonstrate the skills using animation. However, ETS acknowledges this preparation may be inadequate, as candidates are free to review tutorials at any time during the test by clicking on a *help* icon. Unfortunately for hesitant candidates, the test-clock does not pause if this function is activated.

The difficulty is compounded by the typical profile of a TOEFL candidate in Japan. We have seen that the majority of Japanese TOEFL candidates are undergraduates, while those in other countries are typically graduates. Naturally, undergraduate candidates tend to be less familiar with the academic use of computers than graduates; they are also less mature or confident overall.

Finally, a more general caveat may be appropriate. Teachers and researchers occasionally express concern that the growing tendency in education and other fields to *replace* people with computers may ultimately contribute to a subtly "dehumanising" influence (Lewis et al., 1999, pp. 196-7). This is hardly true of the current CBT, as the former TOEFL did not involve human interaction and was already scored by computer anyway; in fact, the formerly occasional Test of Written English (TWE, see below) is now a standard component of the CBT, but is still marked separately by hand, thus introducing an element of human interaction. Nevertheless, ETS apparently has plans eventually to incorporate the former Test of Spoken English (TSE, see below) into the CBT, a development that might more appropriately raise the spectre of "computer-creep."

5. Via the back door: Changes in test philosophy and materials

Ironically, the most far-reaching differences between the TOEFL and the CBT have nothing whatsoever to do with the fact that the TOEFL has been computerised. ETS envisages the CBT as a work in progress, not a finished product. "Eventually, the TOEFL will focus exclusively on English for academic purposes" (Douglas, in Kruse & Shaver, 1998, p. 19). Some types of question formerly common in the paper-and-pencil TOEFL will gradually be phased out in favour of more purely "academic" types (Kruse & Shaver, 1999, p. 75). This shift in emphasis will probably be most evident in Part B of the listening section of the CBT, formerly comprising longer conversations and excerpts from lectures, now comprising discussions and excerpts from lectures. Examples of questions to be phased out apparently include restaurant or café dialogues, speeches by tour-guides, conversations between students, announcements by, or conversations with, university administrators (such as librarians or accommodation officers), doctor-patient dialogues, etc. These are to be replaced by more of the professor's lecture/classroom discussion types of question, also used as test material at present. It is clear this trend in the development of the CBT is already underway. Questions based on material involving two or more students debating a topic with their teacher were never included in the paper-and-pencil test, but are frequent in the current CBT, while long conversations between two students discussing anything other than an academic project have all but disappeared.

Not only have the Part B long listenings become more pedagogical, they have also become more difficult. Candidates for the TOEFL have always been required to answer questions about what they have heard only *after* the target listening passage is over. They have never been allowed to take notes. However, formerly, TOEFL candidates had at least the opportunity to read some of the questions (and accompanying answer choices) in their test books while they listened – an invaluable crutch. In the CBT, however, questions do not appear on the screen at all until after the talk is finished. As these talks are frequently several minutes long and technical in nature, the demand on candidate memory is considerable.

It is not easy to see why ETS has chosen to place such a burden on memory, although one consideration may be a perceived threat to test security posed by, for example, students taking notes. Nevertheless, ETS clearly acknowledges there to be a burden since the linguistic processing involved in Part B of the listening section is usually less sophisticated than that involved in Part A, where conversations are much briefer. Presumably, this is to compensate candidates for the extra demands made on memory. Yet, we may question the wisdom of favouring memory over linguistic skills in a language test. It might be countered that cognition is not so easily compartmentalised, and memory not merely a store of passive data but a creative process intimately involving language (Bartlett, 1932). However, this does not account for ETS choosing to expose CBT candidates to an experience so *different* from what they would meet as students at an American university. It would be entirely natural – indeed, *desirable* – for students on a conventional U.S. college or university course to take notes. It is difficult to understand why a legitimate writing skill, and one which candidates will certainly have to learn if they are to succeed as students in the United States, is effectively discriminated against in the CBT. In this case, computerising the TOEFL has apparently served as an opportunity to make the test pointlessly difficult and unnatural.

6. The Test of Written English

Consistent with this shift in favour of more purely academic language skills is what is probably the most daunting change for Japanese candidates. Formerly, the TWE, a brief essay task, was given at occasional sittings of the TOEFL, and used almost exclusively by ETS to ensure that multiple-choice TOEFL scores were valid guides to candidates' academic writing skills. However, some university departments, particularly postgraduate ones or those (such as literature or history) where academic writing plays an intensive role, sometimes asked candidates to submit a satisfactory essay grade in addition to a conventional TOEFL score. In response to such requests from American faculty, ETS has incorporated the TWE in the CBT as standard; it now accounts for 16.5% of a candidate's total TOEFL score.

This is a challenging development for Japanese candidates for a number of reasons. First, there is the obvious textual difficulty that Japanese learners of English must overcome in writing the Roman alphabet. Second, they are handicapped by having been persuaded, from a very early age, that their very thought processes are so distinct from what are supposed to be a Westerner's that the differences between Western languages and Japanese exist precisely for this reason (see, for example, Suzuki, 1986, p. 142). Third, they must bridge the gulf separating the rhetorical and stylistic traditions of written Japanese and English. Fourth, Japanese candidates do not usually have much experience of writing formal essays even in their own language. Japanese schoolchildren are rarely *examined* on their ability to write more than sentences or phrases in their own, or a foreign, language. Only when they enter university will they regularly be expected to write a number of formal essays. Finally, it is ironic that, although Japanese students write few essays at school, a form of essay endures as a popular genre in modern Japanese literature. Best-seller Yoshimoto Banana, for example, known to the West through translations of her novels and short stories, is also celebrated in Japan for her essays (Sheriff, 1999, p. 279). Young Japanese, plunged into the world of English academic writing, are unfamiliar with the art of formal composition even in their own language, but are influenced by writing models derived from Japanese pop culture.

As with other changes, ETS seems to have given thought to the extra burdens placed upon candidates by the incorporation of the TWE. For the time being, candidates may choose to complete the TWE by hand if they so wish. However, ETS plans to withdraw this option eventually, by which time all candidates will have to compose and submit their essays on computer.

7. CBT – the future of English testing?

The introduction of the CBT was not planned as an experiment, but as the first stage in a natural progression (Kruse & Shaver, 1999, p. 75). If we agree with Gates that in the future, "our computers will be our telephones, our post offices, our library, and our banks" (1995, p. 1), then by extension we may also expect EFL assessment to be available by computer. The question is, how much difference will it really make? In the case of the current CBT, one is obliged to suggest, "Not much!" No doubt we are on the threshold of an age when computer literacy will be paramount, which ETS is wise to anticipate by offering TOEFL on computer.

Nonetheless, it would be premature to celebrate the CBT as heralding a revolution in English testing. While the CBT is in some ways more convenient for candidates than the former paper-and-pencil TOEFL, many changes arising from the process of computerisation are essentially cosmetic. As we have seen, the most significant changes from a candidate's point of view – such as the permanent incorporation of the TWE – are not technological. In seeking to train students for the rigours of the CBT, which is so much more demanding than the former paper-and-pencil test, teachers in Japan will be struggling with already overstretched TOEFL preparation timetables. Nonetheless, were the CBT a less weighty project, one would be tempted to describe the computerisation of the TOEFL as something of a gimmick.

It will be interesting to see if and when other English language tests follow ETS's lead. The SAT and GRE, not strictly designed for non-native speakers, have been available on computer, incorporating CAT, for some time. TOEIC, with its TOEFL inspired multiple-choice format, would be another natural candidate. On the other hand, the production-heavy University of Cambridge Local Examinations Syndicate exam suite (KET, PET, FCE, CAE, and CPE) and IELTS, would presumably be impossible to adapt for computer in any other than the most cosmetic way at present.

8. The Test of Spoken English

As mentioned above, ETS regards the CBT as a work in progress. Among other changes apparently in the pipeline is the incorporation of the TSE as standard in all test sittings (Kruse & Shaver, 1999, p. 75). As it currently requires human interlocutors/assessors, the TSE was formerly only available at certain sittings by prior appointment. In the future, it is envisaged that this be conducted by computer, as part of a conventional CBT battery. How this is to be realised is not clear, and the questions raised are most intriguing. With the reader's indulgence, the author shall speculate on the possible form of a TSE incorporated into the CBT, for it represents one of the most exciting challenges to the CAT ideal. The development of a pure computer version of the TSE would require a substantial *qualitative* extension of the software's adaptive capacity; at least in the short term, ETS will probably be obliged to choose a compromise solution.

On the one hand, it may be intended that the software itself play the roles of interlocutor and assessor; this would be an extraordinarily ambitious undertaking given the present state of natural language processing (nlp) technology. First, the technology would have to incorporate a speech recognition system (SR) able to recognise English language items uttered in a huge variety of non-native accents; it would be unrealistic and unfair to expect candidates to demonstrate "perfect" pronunciation, even in an exam. Second, the technology would have to incorporate a speech understanding system (SU) capable of sustaining a semblance of *purposive human conversation at US undergraduate level*. Presumably, this would require a capacity for reasoning about specific domains, improvisation on the basis of scripts, and a degree of sheer open-endedness (I use this expression as a convenient term for the combination of spontaneity, unpredictability, and freedom associated with natural conversation). The classic but embryonic SU models for corresponding software would be Winograd's SHRDLU (1972), Schank and Abelson's Scripts (1977), and Weizenbaum's ELIZA (1976), respectively (for discussions of ELIZA in the context of CALL, see also Kruse (1998), and – especially for ELIZA's most recent incarnations – Berberich, this volume). However, a perusal of these three SU models amply confirms the judgement of Berberich (Lewis et al., 1999, p. 195) that "the orders of magnitude in human and machine language processing are vastly different, and this suggests that fully conversational CALL is a long way off." Third, the technology would have to have the capacity to *assess* the candidate's spoken performance, in which case the demands on the software would truly be prodigious. The software would have to weigh the semantic content of a candidate's utterance, its grammatical correctness, and appropriacy of vocabulary and register, together with the candidate's spoken discourse skills and accuracy of pronunciation. As there are a vast number of ways in which one utterance may legitimately be expressed, the sheer open-endedness involved places the task of assessment well beyond the bounds of any nlp software known to this author. Some commercial EFL software does assess users' spoken language production by computer quite successfully (for a discussion of some examples, see Berberich, this volume). However, the target language is very simple and severely restricted in semantic scope, and the assessment involves all-or-nothing criteria: if the user's utterance fails to meet a certain threshold of accuracy, it is judged incorrect. This would not be nearly adequate for the TSE, which would have to be able to assess an utterance's *degree relative* to a target threshold of accuracy. Finally, as there would be many types of target threshold to consider, the micro-assessments of the relative accuracies of grammar, vocabulary, pronunciation, style, etc., would each have to be given a due weight and then all combined into one global judgement.

On the other hand, it may be planned to provide a simple audiovisual link from the test-booth to a human interlocutor/assessor, or interlocutor and assessor, elsewhere. This comparatively simple act of *computerisation* might be prohibitively expensive, as it would entail a large number of qualified human examiners being available to conduct interviews at varying times. Nevertheless, in the short term, this would be much easier than developing the software required to replace it. As an artefact, an additional virtue of this approach would be its exploitation of the expanding range of human interaction via electronic media. It is already the case that universities in Europe and the US, and now also in Japan, are making lectures and classes available on television and over the Internet. Frequent so-called *tele-conferencing* for educational purposes is probably not far behind. Kluge (this volume) and Thornton & Houser (this volume) argue that teachers should be making more of the ubiquitous tendency of students to communicate by text-messaging on mobile-phones. As the first generation of mobile phones incorporating video is already on the market, we may anticipate that this also will become a commonplace of communication.

Finally, ETS may opt for a combination of computer-as-interlocutor and human-as-assessor in the TSE. This would still pose a major challenge to software development, but is at least more immediately feasible than computer-as-assessor, given the current stages of development of the various technologies required.

In any case, nlp will have to advance considerably if computer versions of EFL tests are ever to be more than a chimera, a word that the *Concise Oxford dictionary* defines as both a "thing of hybrid character" and "fanciful conception." The first definition admirably describes the CBT at present; the second, what we may realistically expect is the immediate future for the goal of wholly computerised, comprehensive EFL testing. We may be more optimistic in expectations of the distant future. Berberich (this volume) observes that non-adaptive systems are *deterministic*, that is, all possible outcomes are programmed beforehand; while adaptive systems can be said to be *theory-free*, that is, they are self-organising in response to variations in input. In effect, to a greater or lesser degree according to their level of sophistication, adaptive systems *learn*. Present-day CATs are primitive exemplars, but the potential for adaptive systems is enormous. Until recently, artificial intelligence (AI) has been posited on the principle that a non-adaptive, deterministic system could perform the same mental functions as the human brain, were its programming only sufficiently comprehensive, and its processor sufficiently fast. The *locus classicus* for this view is generally held to be Turing (1950). Over a half-century has passed since Turing proposed his test, but nothing like the level of sophistication he predicted has been achieved; Blay (1986) has suggested that Turing inadvertently led AI up a blind alley. Since the human brain is clearly an adaptive system *par excellence*, more recent developments in AI have tended to reflect the view that brain-like capabilities may only be possible in systems that develop them in the same way the brain does: gradually, through learning. A dramatic example of a working computer model of language development of this type is "Hal," named after the (fictional) talking computer in Kubrick's film, *2001: A Space Odyssey*. Hal's designers claim that Hal is "a computer program based on a set of behavioural algorithms that enable the computer to learn language the same way humans do" (Chen, 2001). A neurolinguist spends several hours a day "talking" to Hal, and after no more than a year, Hal was already able to fool child-language experts that he (as his "mummy" calls him) was actually a fifteen-month old child. Hal appears to be learning quickly, and the CEO of the company that developed him predicts that "within ten years, we will see . . . computers that can converse naturally with humans" (Chen). This prediction seems overly optimistic, but Hal's example does suggest one possible road ahead for the development of fully independent, comprehensive, computerised EFL testing: the software might be *trained*. Undoubtedly this is still a long way off, and Hal will have to grow up first. For the time being we must leave the last word to Hal's fictional namesake in *2001*, who, when ordered by his mission commander to open the pod-bay doors, replies, "I'm sorry, Dave. I'm afraid I can't do that."

References

Bartlett, F. C. (1932). *Remembering*. Cambridge: CUP.

Blay, W. (1996). The Turing test – AI's biggest blind alley. In P. Millican & A. Clark (Eds.), *Computers and thought*. Oxford: Clarendon Press.

Carlson, R. D. (1994). Computer adaptive testing: A shift in the evaluation paradigm. *Journal of Educational Technology Systems, 22*(3), 213-224.

Chen, J. (2001). Computer babbles like a baby. *BBC News*. Retrieved February 17, 2002, from news.bbc.co.uk/hi/english/in_depth/sci_tech/2001/ artificial_intelligence/newsid_1537000/1537842.stm

Dunkel, P. A. (1999). considerations in developing and using computer adaptive tests to assess second language proficiency [Electronic version]. *Language Learning & Technology, 2*(2), 77-93.

Gates, W. (1995). *The road ahead*. London: Viking.

Kruse, M. D. (1998). New wine in old bottles: Is there a future for ELIZA after all? In P. N. D. Lewis (Ed.), *Teachers, learners, and computers: Exploring relationships in CALL* (pp. 205-214). Nagoya: Chubu Nihon Kyouiku Bunkakai.

Kruse, M. D., & Shaver, T. B. (1998). Dan Douglas, distinguished visiting professor: An interview. *Journal of Educational Systems and Technologies, 8*, 17-19.

Kruse, M. D., & Shaver, T. B. (1999). Preparing students for the computer TOEFL. In P. N. D. Lewis (Ed.), *Calling Asia* (pp. 73-75). Nagoya: Chubu Nihon Kyouiku Bunkakai.

Lewis, P. N. D., Berberich, F., Adamson, C., Kruse, M., & Ryan, K. (1999). Natural language processing in CALL. In P. N. D. Lewis (Ed.), *Calling Asia* (pp. 193-198). Nagoya: Chubu Nihon Kyouiku Bunkakai.

Madsen, H. (1986). Evaluating a computer-adaptive ESL placement test. *CALICO Journal, 4*(2), 41-50.

Morrison, R. (1999). A new challenge: Preparing students for the computer-based TOEFL. *Journal of Educational Systems and Technologies, 9*, 27-33.

Pelgrum, W. J., & Plomp, T. (1993). The worldwide use of computers: A description of main trends. *Computer Education, 20*(4), 323-332.

Schank, R. C., & Abelson, R. P. (1977). *Scripts, plans, goals and understanding*.

Hillsdale, New Jersey: Lawrence Erlbaum.

Sheriff, A. (1999). Japanese without apology: Yoshimoto Banana and healing. In S. Snyder & P. Gabriel (Eds.), *Oe and beyond: Fiction in contemporary Japan* (pp. 278-301). Honolulu: University of Hawaii Press.

Stocking, M. L., & Swanson, L. (1993). A method for severely constrained item selection in adaptive testing. *Applied Psychological Measurement, 17*(3), 277-292.

Stocking, M. L., & Swanson, L. (1998). Optimal design of item banks for computerized adaptive tests. *Applied Psychological Measurement, 22*(3), 271-279.

Suzuki, T. (1986). Language and behavior in Japan: The conceptualization of personal relations. In T. S. Lebra & W. P. Lebra (Eds.), *Japanese culture and behavior: Selected readings* (pp. 142-157). Honolulu: University of Hawaii Press.

Swanson, L., & Stocking, M. L. (1993). A model and heuristic for solving very large item selection problems. *Applied Psychological Measurement, 17*(2), 151-166.

Turing, A. M. (1950). Computing machinery and intelligence. Reprinted (1990) in M. A. Boden (Ed.), *The philosophy of artificial intelligence* (pp. 40-66). Oxford: Oxford University Press.

Wainer, H. (1990). *Computer adaptive testing: A primer*. Hillsdale, NJ: Lawrence Erlbaum.

Weizenbaum, J. (1976). *Computer power and human reason*. San Francisco: W. H. Freeman.

Winograd, T. (1972). *Understanding natural language*. New York: Academic Press.

16

M-Learning: Learning in Transit

Patricia Thornton & Chris Houser
Kinjo Gakuin University

1. Introduction

Mobile learning (m-learning) is education that is enhanced with handheld mobile devices such as personal digital assistants (PDAs), mobile phones, and eBook readers. These are additional tools that educators and instructional designers have at their disposal as they consider the most effective ways to connect learners with content, with other learners, and with experts or mentors. Like any instructional tool, handheld devices have special features that make them more or less appropriate in each learning situation.

Currently only a few educational projects involve mobile phones. Almost all modern Japanese mobiles can access internet mail and Web; perhaps the best known are the so-called i-mode phones. A few projects in Japan use these and other Internet phones for educational purposes, and a few projects use similar phones in other parts of the world. These projects are experimenting with both administrative functions such as scheduling and notifying students about changes in a syllabus, and learning functions such as the use of mobile phones' short messaging service (SMS) for learning foreign language vocabulary and the delivery of course content.

In North and South America, educational projects involving PDAs are underway. According to industry promotional materials (*Pen World*, 2001), the two types of mobile device, phones and PDAs, are merging, promising an integrated system combining the superior input, computation, and display functions of PDAs with the communication functions of mobile phones. In Japan, PDAs have been less popular than mobile phones, but industry watchers predict an upsurge in that market, due to future capabilities for viewing and sending moving pictures via PDAs (*PDA Market*, 2001) and a new Sharp/DoCoMo PDA with third-generation (3G) wireless capabilities (Sharp, 2001). Thus, this paper will investigate the use of both types of mobile device in educational contexts, presenting an overview of learning with mobile devices. We will analyse the features of mobile devices, relating them to instructional methods and sample projects.

2. Mobile phones in Japan

In Japan, 70% of the population has access to mobile phones, compared with 18% for PCs (Chandler, 2000). Compared with other parts of the world, mobile phone technology in Japan is developing rapidly and the cost structure for its usage is more favourable. Thus, the exploration of this resource as an instructional tool is especially critical in the Japanese context.

2.1 Features of mobile phones
Today's mobile phones can do much more than just connect people. Many mobile phones have personal information management features such as storage and retrieval of phone numbers, calendar and scheduling functions, and sending and receiving email. Newer models can access webpages designed for mobile phones, but this service is still very limited.

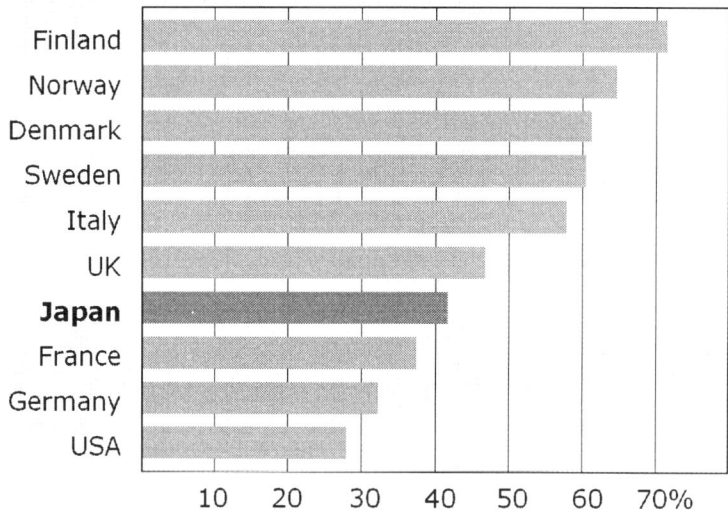

Figure 1. Percentage of total population carrying mobile phones, by country, as of March 2000 (Maitai, 2000).

2.2.1 Penetration
Some 50 million Japanese, about 40% of the population, carry mobile phones. As

seen in Figure 1, this number is less than many other countries, ranking seventh in the world. However, the percentage is much higher when the population is restricted to college students. Informal surveys conducted at several colleges in Japan among students of different ages and academic majors indicate that at least 95% of Japanese college students constantly carry mobile phones. Regarding the introduction of i-mode phones (and similar mobile phones with Internet access) in Japan, Emerson (2001, p. 24) stated, "An instant sensation among teens, it (DoCoMo, a Japanese mobile phone service provider) is still signing up 32,500 customers a day." Also, in the same article, DoCoMo is named as the world leader in data transmission (p. 24), indicating that the services are being accessed frequently. Tertiary-level educators in Japan can assume that most of their students carry and often use some kind of mobile phone.

2.1.2 Bundling

In Japan, mobile phone hardware is sold *bundled* with service plans (network and billing). Thus, phone service providers (e.g., NTT DoCoMo) also provide hardware (manufactured by companies such as Sony) when customers sign up for service. In this sense, each phone is tied to a specific service; it is impossible to buy phones without service, and a phone sold with one service will work with no others. (Most other countries have antitrust laws forbidding such tie-ins.) The Japanese tie-in means that service providers can subsidize hardware. Phones are provided at far below cost, often free, when customers sign contracts for six months' service. As a result, many Japanese perceive hardware as free and frequently trade in their phones for newer models. On average, Japanese trade up their hardware every 15 months (Shah, 2000). This means that most Japanese carry fairly up-to-date phones, and new mobile technology can be introduced and spread rapidly. (This contrasts with other countries, where phones can cost hundreds of dollars, and are updated only every five years.) Educators in Japan can assume their students will have recent hardware, and, in particular, will carry mobile "Internet phones" that can surf the Web and exchange email.

2.2 Problems with mobile phones

2.2.1 Input-output

Although Japanese mobile phones can access standard webpages and email messages, they have restricted input and output interfaces. First, input is slow. English can be input on mobile phones at an average of six words per minute (wpm) (Silfverberg, 2000). (Compare this with an average of 15 wpm on a Palm, and 60 wpm on a PC.) Educators must require little typing on mobile phones, or be prepared for long waits and frustrated students.

2.2.2 Screen size

The display is certainly small. However, several informal investigations have independently concluded that the tiny screen is not a problem for some types of activities. Users of even a four-line display were not bothered by its tiny size for text-based activities (Ring, 2001), and almost all Japanese mobile phones have comparatively generous ten-line screens. Thus, despite the perception of the small screen as a disadvantage, educators can create materials that are effective on diminutive mobile displays.

2.2.3 Media

Network speed is very limited, constraining the types of media that can be effectively used. Almost all Japanese mobile phones connect to the Internet at nine kbps – over five times slower than even the slowest modem on a modern PC connected to a standard phone line. At this speed, Japanese mobile phones can display text, but not graphics, music, or movies. Figure 2 compares the bandwidth provided by various networks with that required for various media. All media are assumed to be sized to fit the displays of mobile phones; static media download occurs in one second.

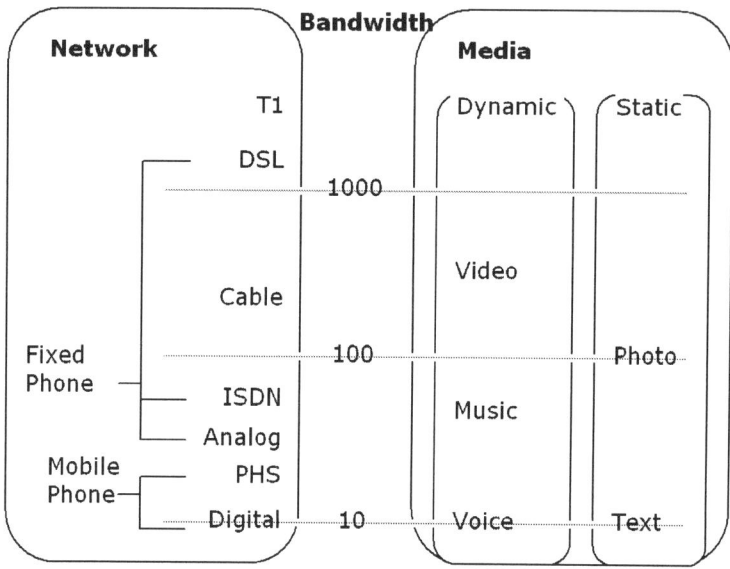

Figure 2. Bandwidth provided by various networks, and required for various media.

Figure 2 shows that modern mobile networks have bandwidth sufficient for only text and low-quality voice media. Even the fastest *PHS* networks are far too slow to transmit photos or stream music and video. Promised 3G (115 kpbs) networks might be able to stream high-quality voice and music, but still not video. For now, mobile educators need to confine themselves to voice with text-only email and webpages.

3. PDAs in Japan

A PDA is a tiny handheld computer. Although not as popular in Japan as mobile phones, makers and distributors of PDAs are trying to set them apart in the minds of consumers by enhancing media capabilities. With these enhancements and with the current software applications, PDAs can become attractive potential educational tools in Japan.

3.1 Features of PDAs

PDAs come in many shapes and sizes. Hardware styles include computer-enhanced watches, eyeglasses, pocket-sized pen-operated units, and larger keyboard-controlled clamshells. There are similarly several different operating systems, but one, PalmOS, dominates the market. Software varies between models, but always includes as a minimum the ability to manage personal information such as schedules, telephone numbers, shopping lists, and memos. Basic models are computerized personal data managers. However, because PDAs are computers, they can also include additional software applications. These more updated models function more like a desktop computer than a simple data manager. Standard software allows PDAs to send email, surf the Web, process words and numbers, and read electronic books (eBooks). PDA hardware, while diminutive, sometimes exceeds the functionality of desktop computers. Higher-end models include wireless modems, wireless LANs, music and video players, and one-button voice memos.

3.1.1 Penetration
PDAs are less popular than mobile phones in Japan. There were an estimated 1.3 million users in 2000, compared with 50 million mobile phone users. Japan's PDA market is expected to triple by 2004 to approximately 4 million users (Makers Trying, 2001).

3.2 Problems with PDAs
3.2.1 Media access
While PDAs can access most of the same documents as desktop PCs, they are rather more limited. For example, PDAs can surf the Web and view text and pictures, but lack the sundry plug-ins required to view less common formats, such as QuickTime and Real Video. Also, modern PDAs can edit *some* information in files created by Microsoft Office, but lack many features of Office itself. PDAs often have slower processors and networking connections, and so fail to run some games and streaming media. Of course, new software constantly increases the capabilities of PDAs, but in some aspects they will always be behind desktops.

3.2.2 Text entry
Although faster than mobile phones, data entry on PDAs is still slow compared with desktop computers. When using a pen-like stylus, some PDAs require the user to learn a new way of printing the alphabet for English input. For Japanese input, the pen can be used to write a Japanese character, but the quality of character recognition varies between models. Attachable keyboards can solve some of these problems, but trade entry speed for portability.

4. Mobile devices in education

As we begin to explore m-learning, we must ask in what ways mobile devices can adequately support educational endeavours. Education involves a variety of activities such as content delivery or access, assessment, communication between learners and experts, communication between learners, interaction with various forms of information, and management. Below we consider some recent projects involving mobile devices and the educational functions they facilitate.

4.1 Mobile devices and content

4.1.1 eBusiness on the move: Content via WAP.

In Singapore, an eBusiness course used a combination of Web and Wireless Application Protocol (WAP – miniature Web for mobile phones) technologies to deliver the
entire course (Ring, 2001). There were no face-to-face meetings. Seventy percent of
the course was available in both modes, including all text content, quizzes, and schedules. Ten percent of the course used only WAP phones (provided by Nokia) to access
WAP sites and for quick alerts or reminders from coaches and instructors. Twenty
percent was available only on the Web, including digital video clips, bulletin board
discussions, PDF files, and email exchanges. Thus, a total of 80% of course activities
could be accessed via mobile phones.

There were some special design considerations for parts of the course available in
both Web and WAP modes. The two technologies required different organization of
content. Because of the smaller screen, the WAP course material had to use shorter
chunks of text, more screen displays, and more section titles. For example:

Web Files/Headings	WAP Files/Headings
eBusiness on the Move	eBus on Move
Master Guide	Master Guide
1. Course Welcome	Welcome
2. Course Overview	1. Introduction
	2. Coach Tung
	3. Coach Angehrn
	4. eLearning
	5. Getting info
	6. Personal info
	7. Knowing others

Thus, what was one click on a PC became eight tiny pages on a mobile phone. The
different organization of the two modes was a concern for the designers who were
afraid that learners would be unable to switch easily between them. This problem was
solved by sending a Word document, which cross-referenced the two modes, to all
participants of the course.

At the end of the course, participants evaluated the use of mobile phones for learning. One hundred percent of respondents agreed that having WAP access added value
to the learning experience, commenting on increased usefulness, immediate feedback, and ability to study texts while commuting using the mobile phone. Ninety
percent agreed that having WAP access made taking the course more convenient.
When asked about the feasibility of delivering an entire course by WAP alone, there
was a mixed reaction. Approximately half the group were doubtful, citing WAP's
small screen size, lack of graphics, and limited interactivity.

Although doubtful about a full course via mobile phone, many learners, surprisingly, found the small page size of their phones actually *advantageous* for text activities. This size restriction forced authors to decompose text into smaller modules, which
turned out to be easily digested, even in the often-distracting mobile environment
(e.g., reading while commuting on public transportation). Furthermore, the WAP index to these modules was necessarily a deeply-nested menu of links. Authors feared

this menu would intimidate students, but in fact the students found the detailed menu a helpful outline of the course, and found navigating through this large structure a good way to gain an overview and sense of the scope of the course. They considered such a detailed outline much more useful than the coarse-grained table of contents found in most printed books and websites, which list only a small number of large chapters, or long webpages, and so present only a vague overview of the content. Surprisingly, the WAP technology forced a more usable interface.

This project showcases the use of mobile phones to enhance Web-based courses. Educational activities delivered via mobile phone included content delivery, assessment, management tasks, and communication tasks. These activities used both the *pull* and *push* aspects of mobile phone technology with students having access to course content at their convenience but receiving pushed reminders and alerts from coaches. In this project, WAP sites were created, identical in content but different in organization to websites for the same course. In addition, phones were used to take quizzes, check scheduling, receive reminders from instructors, and to talk to other students on the course. From this project, it seems that current mobile phones with WAP can deliver textual content effectively by decomposing the text into small modules, but large files, such as PDF and video, were available only for PCs. Mobile phones can be an useful addition to Web-based courses.

4.1.2 Learning on the move: Content via SMS

In Japan, a research project (Thornton & Houser, 2001; Houser, Thornton, Yokoi, & Yasuda, 2001) was conducted using SMS on mobile phones to deliver English language vocabulary instruction to 44 female Japanese students enrolled in at least one of two courses in the Department of Language and Culture at Kinjo Gakuin University. Based upon research on memory and learning (e.g., Dempster, 1987, 1996; Bahrick, 1987) the focus of the study was to provide a method of vocabulary learning including rehearsal at timed intervals. Mobile phone SMS was chosen as a push media that would promote regular study. Both usability and learning issues were explored.

Students received short mini-lessons (less than 100 words of text) three times a day, providing opportunities for learning and practising five to seven words per week. Mini-lessons included definitions of single words, multiple uses of a word in context, review of previously introduced vocabulary, and story episodes that incorporated target words. Pre- and posttests determined the number of words learned during each two-week cycle.

Usability results

Students were asked to evaluate this mode of learning by responding to a questionnaire. Seventy-one percent of subjects reported preferring mobile phones to PCs. Ninety-three percent responded positively when asked, "Is this a valuable teaching method?" Also, eighty-nine percent reported a desire to continue learning via phone. Sixty-nine percent indicated that the small screen size was not a problem.

Learning results

Regarding the promotion of regular, interval study, forty-six percent reported reading messages three times per day, thirty-four percent once a day, and twelve percent twice a day. Although a majority of students read lessons at least once per day, the varied frequency indicated limited success in promoting carefully timed interval study. In both classes, posttest results showed that SMS lessons promoted learning, even without any outside motivation for study (i.e., course grades) (Class 1: mean pretest score=2.0, posttest=4.6, p<0.05; Class 2: mean pretest score=2.26, posttest=4.41, p<0.05). When the same lessons were given via Web (pull media) and SMS (push media), the results showed an average of three words learned via Web, compared with an average of 6.5 via SMS (p<0.05). Comparing lessons delivered via paper (pull media) and SMS, results showed that only forty-eight percent of students using paper-based lessons improved their score on a posttest, where eighty-eight percent improved via SMS. In one class, students were given extrinsic motivation for study when the posttest grade counted as a quiz grade for the class. During this cycle, the average increase in number of correct responses on the posttest was seven, compared with earlier (non-graded) results of three and four.

Future phases are planned that will continue studying learning and usability issues. Variables in content such as use of first language definitions, glossing of target words, and length of messages will be studied. Variables in delivery, including the number of times a word is repeated and the number of messages received per day, will also be considered. In addition, future studies will investigate interactivity with Web-based quizzes and feedback accessed via mobile phones.

This project, utilizing mobile phones to deliver content via short messaging service, shows that the delivery of foreign language vocabulary lessons via mobile phones is effective and received positively by Japanese university students. Vocabulary lessons are easy to break into discrete chunks that can be sent and read via mobile phones. This method of delivery, taking advantage of the push aspect of mobile technology, can promote a more regular approach to the study of vocabulary, though the exact timing of intervals cannot be controlled since students have freedom to choose when each message will be read.

4.1.3 Mobile learning at Stanford Learning Lab: Content via audio

The Stanford Learning Lab (Regan, Mabogunje, Nash, & Licata, 2000) investigated the use of voice and email on mobile phones for filling in gaps of daily time with learning opportunities. They developed a few rough prototypes that could be used for foreign language (Spanish) study via mobile phones. The prototypes included practising new words, taking quizzes, accessing word and phrase translations, talking with a human coach, and saving new vocabulary to a notebook. These activities formed an integrated voice/data environment.

The purpose of this study was to gather information that might be used for future research. Only informal tests were used to evaluate the effectiveness of prototypes and user opinions. This project used mobile phones to connect teachers and learners, access content sources, and do informal assessments. Their general conclusions were that mobile phones were effective for delivering quizzes due to the small question chunks accessible in available bits of time. They thought that the small screen size made this technology unsuitable for learning new content but effective for review and practice. The live coaching sessions over mobile telephones were perceived as effective for language learning and practice. However, phone connection quality was a problem for comprehension and the amount of time spent was difficult to manage.

The final prototype involved automated voice-controlled vocabulary and quiz sessions. The researchers felt that this mode of delivering audio has great potential. At this time, the technological challenges for provision via mobile phones are still too great to make it practical. However, in the future, personalized, database-driven listening and speaking practice could become a reality. These researchers advised others interested in m-learning to focus on the parts of the learning process most suited to audio, small chunks of time, and a distracting environment.

4.2 Mobile devices and quizzes

Two of the projects described above – eBusiness on the Move and the Stanford Mobile Learning project – incorporated WAP or text quizzes as part of their mobile learning program. In both cases, participants were satisfied with the format and usability of quizzes on mobile phones, and multiple-choice questions were easily answered using the limited keyboard. Quizzes comprise small question-and-answer modules and are thus suitable for mobile use where modules are easily viewed on small screens, and the time available is short and often unpredictable (e.g., while waiting for a bus).

Other organizations are also making quiz-type learning available on mobile phones. EnglishTown.com (www.englishtown.com), an English language learning website, promises daily WAP "lessons" in short question/answer format. An i-mode site (iaxs.co.jp/pt/lang.html) in Japan has collected a variety of learning opportunities, including medical terminology and English and Japanese language learning. Many utilize a quiz format that challenges users to answer a short question and then reveals the correct response. A Hangman game that fits on the small mobile phone screen is also available at this site for English language learning. For Japanese learners of English, EigoTown.com includes i-mode lessons for learning idioms and a role-playing game, a type of hypertext fiction.

In all these projects, learning opportunities via quizzes and games are provided for people with unoccupied commuting time or who have no access to a desktop PC. This is utilizing the special nature of mobile devices: their mobility and ubiquity.

4.3 Mobile devices and simulations

In North and South America, educators are beginning to use PDAs as inexpensive PCs for the classroom. Unlike mobile phones, additional software can be installed on PDAs, making them useful as tools for educational activities such as simulations and collaborative projects. Here are a few examples that researchers at the Universities of Michigan and Texas have developed.

4.3.1 Cooties: A germ simulation

The simulation begins when the teacher *beams* (transmits by infrared) a message with a possibly infectious "germ" to each student's PDA. Students then walk around the room, meeting each other, and beaming their messages to other students. At the end of the meeting time, the handheld devices have stored each meeting and students are able to analyse and understand how disease spreads (Soloway et al., 2001).

4.3.2 CritterVille: An ecosystem simulation

Each student creates a "critter" (artificial life form) on a PDA. The actions of each critter are then transmitted wirelessly to a website hosting CritterVille, which simulates interactions between critters. Next, the CritterVille website is able to wirelessly transmit the new information to the PDAs. By generalizing from their own experience of how differently designed critters perform differently in social simulations, students learn about the difficulties of surviving in a complex ecosystem (Solway et al., 2001).

4.3.3 Educational games

In Chile, researchers have been developing educational games for PDAs that teach basic maths and language skills. Students work both independently and collaboratively to solve problems. Some games are associated with a story that allows students to roleplay and participate in decision-making (Rodriguez et al., 2001).

In all the above simulations and games, the PDA functions simply as a classroom computer. The same activity could be done as effectively or better with networked PCs. The mobile nature of a PDA is not being utilized. However, the portability and lower cost of handheld devices make them an attractive alternative to PCs.

5. Mobile devices in field work and management

In the business world, the idea of *just in time* (JIT) information is well known. Handheld devices that facilitate this kind of timely information access are used widely to support workers who must be away from their office PCs. In education, there has been less emphasis on the need for JIT information. However, a recent study by researchers at Cornell University has begun to address this issue, looking at the use of handheld devices for science students who must often work in field locations. Rieger and Gay (1997) developed PDA software applications to support the study of plant genetics. Science students often work outside, collecting data and analysing the environment. In the past, they have later transferred their data to the computer or made notes about questions for further investigation in books or databases. With the prototype applications in this study, most such data recording and information access is available in the field. In addition, in-field collaboration between team members can be facilitated with mobile technology. They are currently evaluating student use of these tools.

Although this is a work in progress, developers of the applications have been able to give some guidelines for mobile learning tools. They recommend context-free (modular) information, content specific to a field environment, and a focus on tutorials, data collection, and diagnostics.

In language learning, we have been unable to identify projects that focus on JIT aspects of data retrieval or collaboration. However, in English as a Foreign Language (EFL) settings, a lot of language learning occurs outside a formal classroom, as learners interact with members of their communities. The use of PDAs and mobile phones for immediate access to phrases and translations is an obvious use of mobile tools for language learning. There are a few incomplete examples of bilingual dictionaries in i-mode mobile phone format (iaxs.co.jp/pt/lang.html and www.eigotown.com). More complete versions are already available in the form of electronic dictionaries and phrasebooks, but by incorporating them into a PDA or mobile phone environment, students can potentially have a more integrated system of learning at their fingertips. Some teachers in Japan have limited access to computers in the classroom. At the college level, many teachers are part-time, moving between schools without the benefit of an office. Teachers are "in the field" during those hours away from office or staffroom. For such educators, the use of mobile devices for recording student progress, taking attendance, and accessing JIT information sources could be valuable. For PDAs, there are already grade book applications and other database resources. Internet capable mobile phones could make information access immediate.

More students in Japan own mobile phones than PCs. For JIT scheduling information, syllabus changes, and other time-critical information, mobile phones are a good delivery mechanism. One educator at a university in Fukuoka, Japan (Daniels & Pellowe, 2001) creates dynamic webpages that can be viewed on i-mode phones. These pages give important information about deadlines and preparation information for upcoming class discussions. At Kinjo Gakuin University in Nagoya, students receive information about schedule changes, job-hunting, and future interview opportunities via mobile phone email and WAP. In the eBusiness on the Move project in Singapore, students could register for the course and receive reminders from coaches on WAP phones.

6. Information flow and learning

As seen above, a few projects are beginning to investigate the use of mobile handheld devices for education. Based on what we already know about learning and education and are finding out about this new technology, we need to consider how mobile devices can appropriately assist in the learning process. To what extent can m-learning be applied in today's classes? Almost all classroom environments can be divided into instructivist and constructivist paradigms. In order to see where m-learning can be applied, we need to look at models of each paradigm.

The instructivist approach, based on models of learning from behavioural psychology (Reeves, 2002), breaks knowledge into hierarchies of topics that are studied in a structured way. The teacher's role in this approach involves determining and presenting chunks of knowledge in an appropriate sequence. These chunks are often presented as objective facts to be memorised by the student and later tested. The learner's role is to absorb and process information later reiterated to demonstrate learning. Methods and tools often used to support this approach are lectures, sequentially structured texts and activities, practice drills, and objective tests and quizzes. The information flow in this learning paradigm is reflected in Figure 3.

The primary flows of information, indicated by the darker arrows, are from the instructor and textbooks to the student. A smaller flow "feeds back" from the student to the teacher, allowing the instructor to evaluate learning and teaching.

A constructivist approach, on the other hand, is based on models of learning from cognitive psychology (Fosnot, 1992). This approach demands more autonomy on the part of the student, giving opportunities to explore and situate learning in practical problem solving. The teacher's role in a constructivist learning environment is to identify and present problems, and then coach students as they attempt to construct solutions. Assessment tasks employed, such as portfolios and presentations, are appropriate to a problem-based learning environment. The learner's role is to actively construct knowledge by participating in, and reflecting on, problem-solving activi-

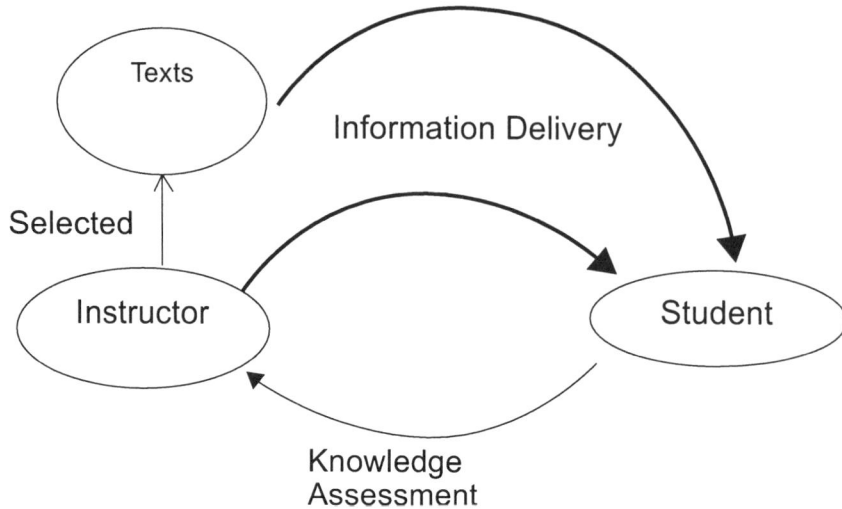

Figure 3. Data flow in Instructivist education.

ties, often as a group member. The learner is responsible for finding and assembling the pieces of knowledge acquired to create a meaningful whole. Methods and tools used in this approach include realistic problems, fieldwork, group projects, self-directed research in libraries and on the Web, and computer-based simulations allowing hypothesis testing. The information flow in a constructivist paradigm can be seen in Figure 4.

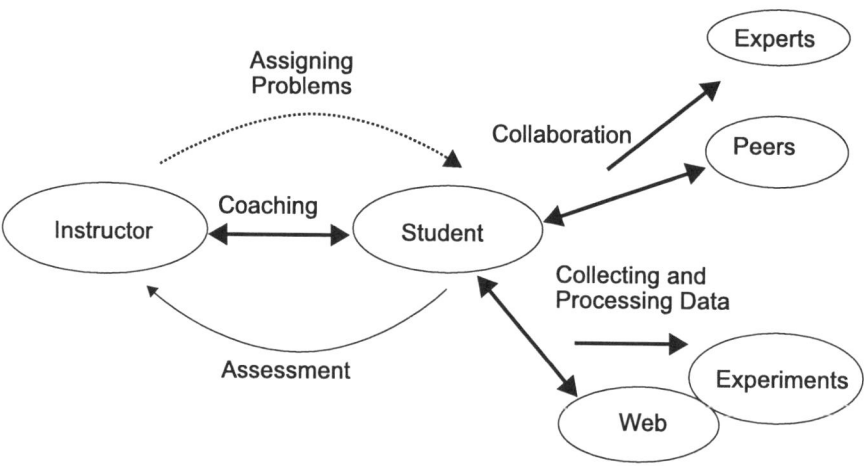

Figure 4. Data flow in Constructivist education.

The major flows of information centre around the student. They interact with peers and with information sources, including the instructor, as they engage in problem solving.

Pure instructivism and pure constructivism are the two extremes. Many instructors use a combination of instructivist and constructivist activities in their courses. Looking at the data flow diagrams above, we can imagine introducing technology by placing it along the lines and arcs of information flow. Thus, instructivist tasks might include using technology to deliver lectures via video or PowerPoint slides on the Web, create structured texts, or evaluate students' progress through objective, automated tests and quizzes. In constructivist tasks, technology could connect learners for collaboration, provide information needed for problem solving, or create Web-based, collaborative projects. Mobile devices, as one type of technology, can support certain activities in both these approaches to learning and teaching. Instructivist tasks developed for mobile devices include structured content delivery, objective quizzes, and foreign language vocabulary and listening activities. Constructivist projects involving mobile devices include simulations, group communication, and fieldwork. Activities not easily supported by most mobile devices at this time include accessing content with complex graphics or sound and the use of highly interactive Web-based activities.

7. Conclusion

Mobile phones and PDAs are tools that can help educators create blended educational opportunities giving learners options about time and place, and thus enhancing other types of delivery methods. At this time, m-learning is constrained by its inability to display large graphics or highly interactive websites. Additional software applications cannot be added to mobile phones. Although all new phones support Java, several problems must be overcome before useful applications can be developed. The Java sites (and graphic format and sizes in general) are nonstandard, and (positional and text) input and (graphic and sound) output are severely limited, so existing applications are ineffective. Most current applications in Japan are games.

Learning activities should be designed that take advantage of the special characteristics of mobile devices and minimize the disadvantages. These include delivering modular chunks of information for just-in-time learning, practice and review of discrete chunks of data, recording, communicating, and accessing information in field settings, and sending reminders or time-critical information to learners.

The few research studies currently underway show some success in m-learning:
- The use of mobile phones as a push media for foreign language vocabulary is effective.
- PDAs can be used to help science learners involved in fieldwork.
- Mobile phones can help learners practice and review small chunks of information in quiz format.
- Modular chunks of content can be effectively accessed via mobile devices.

As the technology of mobile handheld devices improves, there will be a wider range of effective and appropriate uses of handheld devices in education. As Kluge (this volume) points out, students and teachers will no longer be tied to a particular place and time for computer use. Mobile devices with wireless transmission capabilities will make the use of computers – for routine tasks in the classroom and beyond –

ubiquitous. By carefully considering the features and capabilities of the various tools at their disposal and matching them to learning goals, educators can create effective and interesting opportunities for learning. For language teachers, Kluge (this volume) presents ideas for the use of mobile devices in language learning. For other academic disciplines, groups such as the Concord Consortium (www.concord.org) and the University of Michigan's Center for Highly Interactive Computing in Education (hi-ce.eecs.umich.edu) give ideas for the use of mobile devices in the classroom.

References

Bahrick, H. P., & Phelps, E. (1987). Retention of Spanish vocabulary over 8 years. *Journal of Experimental Psychology: Learning, Memory, and Cognition, 13,* 344-349.

Chandler, C. (2000). In Japan, the Internet without the PC [Electronic version]. *Washington Post* (February 8, 2000; Page E01). Retrieved August 21, 2001, from www.washingtonpost.com/wp-srv/WPlate/2000-02/08/0331-020800-idx.html

Daniels, P., & Pellowe, B. (2001, May). *Creating dynamic websites with PHP.* Presentation at JALTCALL2001, Gunma, Japan.

Dempster, F. N. (1987). Effects of variable encoding and spaced presentations on vocabulary learning. *Journal of Educational Psychology, 79,* 162-170.

Dempster, F. N. (1996). Distributing and managing the conditions of encoding and practice. In E. L. Bjork & R. A. Bjork (Eds.), *Memory* (pp. 317-344). Boston: Academic Press.

Fosnot, C. (1992). Constructing constructivism. In Doffy & Jonassen (Eds.), *Constructivism and the technology of instruction: A conversation* (pp. 167-176). Hillsdale, New Jersey: Erlbaum.

Houser, C., Thornton, P., Yokoi, S., & Yasuda, T. (2001). Learning on the move: Vocabulary study via mobile phone email. *ICCE 2001 Proceedings, 1560-1565.*

Maitai, S. (2000). *Picking winners.* Retrieved April 16, 2002, from www.washingtonpost.com/wp-srv/WPlate/2000/02/08/0331-020800-idx.html

Makers trying to set PDAs apart from cell phones. (2001, April 5). *AsiaBizTech.* Retrieved August 21, 2001, from www.nikkeibp.asiabiztech.com

PDA market to expand rapidly, NTT DoCoMo chairman says. (2001, July 24). *AsiaBizTech.* Available from www.nikkeibp.asiabiztech.com

Pen world. (2001, July). *Pen computing.* Available from www.pencomputing.com/features

Reeves, T. (2002). *Evaluating what really matters in computer-based education.* Retrieved from www.educationau.edu.au/archives/cp/reeves.htm

Regan, M., Mabogunje, A., Nash, J., & Licata, D. (2000). *Mobile learning.* Retrieved July 25, 2001, from sll.stanford.edu/projects/mobilelearning

Rieger, R., & Gay, G. (1997). Using mobile computing to enhance field study. *Computer support for collaborative learning '97 Proceedings.* Retrieved August 21, 2001, from www.oise.utoronto.ca/cscl/papers/rieger.pdf

Ring, G. (2001, June). Delivering eLearning using Web and WAP. *ED-Media 2001 Proceedings* (pp. 1567-1572). Norfolk, Virginia: Association for the Advancement of Computing in Education.

Rodriguez, P., Nussbaum, M., Zurita, G., Rosas, R., & Lagos, F. (2001). Personal digital assistants in the classroom: An Experience. *ED-Media 2001 Proceedings* (pp. 1567-1572). Norfolk, VA: Association for the Advancement of Computing in Education.

Shah, S. (2000). RF Requirements for 3G mobile systems. *Compound Semiconductor, 6*(4).

Sharp to make 3G PDAs for NTT DoCoMo. (2001, August 22). *Yahoo Daily News: Reuters.* Retrieved July 25, 2001, from dailynews.yahoo.com/htx/nm/20010822/tc/tech_sharp_pda_dc_1.html

Silfverberg, M., MacKenzie, I. S., & Korhonen, P. (2000). Predicting text entry speed on mobile phones. *Proceedings of the ACM Conference on Human Factors in Computing Systems (CHI 2000),* 9-16.

Soloway, E., Norris, C., Blumenfeld, P., Fishman, B., Krajcik, J., & Marx, R. (2001). Handheld devices are ready-at-hand. *Communications of the ACM, 44*(6), 15-20.

Thornton, P., & Houser, C. (2001). Learning on the move: Foreign language vocabulary via SMS. *ED-Media 2001 Proceedings* (pp.1846-1847). Norfolk, VA: Association for the Advancement of Computing in Education.

17

Tomorrow's CALL:
The Future in Our Hands

David Kluge
Kinjo Gakuin University

1. Introduction

Even among longtime practitioners of CALL, there has been dissatisfaction with the status quo of CALL in Japan. Out of hearing of colleagues, we say to ourselves, "There has to be something better than this!" There is something better, and that "something" may one day be called *MALL* or *HALL*, mobile-assisted or handheld device-assisted language learning, specifically, the preferred handheld device – the new generation of mobile phones. This paper will list problems with current CALL in Japan, describe the new technology, show how it can alleviate some major shortcomings of CALL, and, most importantly, describe possible activities for the language classroom. The paper finally looks at problems of the new technology and pedagogy.

2. Problems with the status quo

In this paper, the status quo – the CALL situation at many institutions in Japan – will be discussed in terms of practical and pedagogical problems.

2.1 Practical problems
Practical problems include computer room scheduling, platform compatibility, the necessity of a hardwired LAN, and cost. Since there is usually one computer room for

many courses, there are often scheduling problems. Even having several computer rooms, with a large student body, a variety of departments, and a varied curriculum, scheduling problems still occur. Looking at the problem another way, the scheduling problems result from the fact that CALL is usually "place locked," meaning one can only use CALL in the computer room.

There are also problems with platform compatibility. Differences between Windows and Macintosh operating systems compound scheduling difficulties and add extra problems. Teachers and students are usually trained only on one platform. Necessary software for a particular course is usually available on only one platform because of the cost of installing software on both.

Schools without a CALL lab have to find resources to build one. This includes purchasing computers, peripherals, software, technical support, and furniture, as well as installing and maintaining a hardwired LAN for networking and Internet access. Sometimes a whole new building must be constructed. There is no certainty that the enormous cost of hardware, software, technical assistance, and construction is recovered by use of the CALL lab.

2.2 Pedagogical problems

Practical drawbacks are many, but more problematic arc pedagogical difficulties with CALL. These pedagogical problems include the steep learning curve, difficulty of assigning homework, perception that computers are only used in the computer room, the stationary nature of computers, the hide-and-seek effect, and the dispersal effect.

The steep learning curve means that much time is spent teaching to use the computer or the software. This is often taken from class-time better used for teaching the skill that is the focus of the class.

Another pedagogical problem with present-day CALL is the difficulty of giving homework, as students may not have the "right" setup at home, and school facilities are unavailable outside class-time due to overscheduling. In addition, computer rooms often close early so it is difficult for students to complete CALL homework.

Students currently feel that CALL is only done in the computer room, in effect psychologically locking CALL into one room, whereas a good system of learning should enable students to use it wherever and whenever they want.

In language teaching today, groupwork is important, with the possibility of the class grouping changing several times during a period from independent work, to pairs, small groups, and the whole class. While possible in the present CALL situation, especially with networked computers, it is awkward. Computers are stationary units that make it difficult for the mobility necessary to create different groupings for different activities.

In addition, typical desktop computers and monitors are bulky, making it hard for teachers to see students' faces and for students to see classmates' faces. When added to the fact that a computer requires considerable space, dispersing students farther from each other, we see that present computer rooms can interfere with the development of class unity. A row and column arrangement with large displays makes it difficult to monitor students. Students play a game of hide-and-seek with the teacher, hiding behind their monitor when they want to avoid the teacher's eye, often surfing the Web or sending email instead of attending to the classroom activity. There are technological ways to overcome this, but because of the "Big Brother" appearance of such technological fixes, like the teacher being able to monitor a computer without the student's knowledge, solutions are often distasteful.

Some schools provide a laptop computer for each student to use at home which eliminates the homework problem, but desktop computers are usually used on campus so the other problems still exist.

3. Why handhelds?

Luckily, instead of just making do with the present CALL situation, or throwing CALL away and going back to pen, paper, and textbooks, there is a good technological alternative: HALL, handheld device-assisted language learning. Why turn to handheld devices? Strong arguments for their use have been made by Soloway et al. (2001) in what may become a seminal article in handheld pedagogy. After looking at surveys of computer use in K-12 schools in the US, they concluded that computers are not being used much. They hypothesize that this is due to problems of distance, time, number, size, and expense. The computer room is usually far from the classroom, adding commuting time. Because of the huge cost per computer, there are often not enough computers for each child, making the name "personal computer" a misnomer, since each is shared. They contend that computers are too big to be used comfortably by young people. In their poetic, prophetic way they state:

> As long as computer labs are down the hallway and up the stairs, teachers will consider them irrelevant to learning and teaching. As long as the ratio of students to computers is 4-7 to 1, the effort needed to use them is simply too high, given all that has to be accomplished in a school day. As long as computers are not ready at hand, they will not be used in a routine, day-in, day-out fashion; the impact of computers on K-12 education will continue to be essentially zero. (Soloway et al., 2001, p. 17)

They reason that, like the computer's successful predecessors, pencil, paper, and calculator, "computers must be within arm's reach, mobile, and palm-accessible in order to make a difference in the classroom," and propose that each student be given a handheld device (Soloway et al., 2001, p. 15). The same arguments hold true for all levels of education in the field of English as a Second Language (ESL).

A description of the technology to be used is necessary before discussing the new pedagogy of HALL. The discussion of the new HALL technology will be divided into hardware, software, and services, considering present and future situations.

4. Present hardware

Handheld devices are often considered to be small computers, looking like miniature laptops barely fitting in the hand like the Sony Giga Pocket, or PDAs (personal digital assistants), sometimes called PIMs (personal information managers), like the Palm Pilot. Two other types of handheld device to consider are mobile phones and game machines like the Pocket Gameboy.

4.1 Handheld computers

Handheld computers look like miniature versions of laptop computers. The main characteristics of palmtop or handheld computers are that all have keyboard input, some have a digital camera, and some have wireless Internet access with the help of cards, while others require a cable to hook up to a mobile phone for Internet access. One such device is NTT DoCoMo's Sigmarion, which runs on Windows CE and has dimensions of 189 x 107 x 27mm at 485 grams. It connects to the Internet via a cable attached to a mobile phone.

Problems

The problem with this type of device is that, while the keyboard is too small for even medium-sized hands, it is not really a handheld device because, although one can hold it, it must be set down on a flat surface for use. In addition, it is as difficult as a computer to learn how to use. Finally, accessing the Internet with one is awkward.

4.2 PIMs and PDAs

In 1996 the Palm Pilot personal digital assistant was introduced, starting a new period in mobility in computing (*The Japan Times*, August 9, 2001, p. 15). PDAs have most often been used for scheduling and as an address book by the business world, and then gradually by the general public. Soloway et al. (2001) have shown that PDAs are a viable alternative to CALL. PDAs have recently become popular in Japan. The Japanese ranking of top PDAs taken from a Japanese handheld device magazine (*K-tai Oh*, pp. 106-111) shows the Sharp Zaurus MI-L1 as the most popular PDA. Other PDAs in Japan, listed in order of their popularity are Sony's Clié PEG-N700C, Palm's m505, Compaq's iPAQ Pocket PC, Springboard's Visor Edge, and HP's jornada 525. (For URLs of these and all other products mentioned in this article, see Appendix B.)

Most of these are small, with a width of 71 to 83.5mm, a height of 114 to 139.5mm, and a depth of 11 to 17mm. They are light, from 136 to 260g, have adequate computing power, from 33 to 206Mhz and capability for text (stylus, thumb keyboard, and keyboard), sound, music, photo, and Internet use when connected to assorted cards and attachments. They can also connect to a desktop computer. The Zaurus can become a phone with proper cards and attachments. The display is small, usually 3.5 inches, but adequate for the popular uses.

Problems

Soloway et al. (2001), although fans of the PDA, also recognize its failings: generally low computational power for multimedia (although newer models like the hp jornada and the iPAQ are more powerful), pen or stylus input which makes extended writing difficult (not a problem if the PDA is connected to a portable keyboard), and a small screen (p. 17). Another problem is that to do many things that are possible with the PDA, one needs to buy various attachments separately. When added to the PDA, it looks like an ad hoc device; one wonders why these attachments were not built in originally. A potential drawback is that accessing the Internet is not a smooth, easy process. Finally, what appears a minor point, but which may have a large impact on the spread of the PDA, is indicated by a quote from actress and designer Sunny Kate: "I was on the set one day, and while looking up something on my planner, I stopped and thought, 'Why is this thing that I use all the time so boring?'" So she designed

PDA holders in a wide variety of fashionable patterns. (*Newsweek*, June 11, 2001, p. 59). The main point is that the PDA in itself, without the optional fancy wrapping is not so stylish, and is unlikely to make inroads into youth culture, and our students.

4.3 Handheld games

Handheld games generally contain little of interest for language teachers, but one feature the games have had for a long time is wireless interactivity with others having the same game set and game. One example of such a game is POX, made by Hasbro, which allows players to interact with each other. A player builds a warrior character on the game set. When two units come into range, the two warriors fight each other. When more than two units come together, they fight each other in a round-robin style (*Newsweek*, July 9, 2001, p. 8). What should be interesting to language teachers is the ability for interactivity among participants. This wireless interactivity involves not only individuals, but groups pairing up, interacting, and moving around, forming new pairs. Of additional interest to teachers is that, unlike POX, most game sets are modular in that software (games) is added through small removable cartridges, and peripherals like digital cameras can be inserted into the same game set slot.

Problems

The problem with game devices is the lack of appropriate or usable software. Game sets are equipped with very limited input devices, usually just a few function and arrow buttons, certainly insufficient for language activities. So far, no game sets are Internet capable, but with the increasing popularity of Internet gaming, this may change. In addition, the design of game sets is for a very young audience.

4.4 Mobile phones

There has been a global boom in mobile phones. First used in business, then by adults in the general public, now even elementary school students carry their own mobile phone. The Japanese – especially the young – have taken to the mobile phone. As of July, 2001, the Telecommunications Carriers Association of Japan stated that population penetration has reached 50.5 percent, with 64.18 million mobile phones now used in Japan (Yoshida, 2001, p. 14.) One mobile phone company, NTT DoCoMo, ranks second in the world in size with about 38 million users, of which about 50 percent are in their teens and twenties. It signs up 32,500 customers a day. However, it ranks first in the world in the amount of data transmission, with second and third place companies very far behind. Thus, Japanese mobile phone owners use them more often than users in other countries (Emerson, 2001, pp. 24-25).

Since mobile phones are the focus of this paper, their discussion will be separated into present hardware, software, services, and problems to be addressed before HALL is adopted by many institutions. Then will come a discussion of envisioned future developments in hardware, software, and services.

In the year 2000, NTT DoCoMo spent nearly $15 billion buying shares in wireless communications companies around the world, with their Japan experience touted as a great success (Emerson, August 6, 2001, p. 24). It will provide KPN Mobile NV of the Netherlands with its special mobile phone technology (*The Japan Times*, August 25, p. 8). It is no wonder that most of the world thinks that NTT DoCoMo is the only mobile phone service in Japan. In fact, there are three main companies (*carriers*) in direct competition with NTT DoCoMo: J-PHONE, au, and TU-KA. The hardware, software, and services of the four carriers are somewhat similar, but there are some unique and potentially useful features of the different carriers, described below.

4.4.1 Present hardware.

In a sense, some of the best hardware in Japan today is future hardware. In May 2001, in a pilot program, NTT DoCoMo unveiled two models of the fastest new generation phones in the world, the FOMA N2001 and the FOMA P2401, capable of sending at 384Kbps, six times faster than other mobile phones in Japan. Both handsets are very stylish. The N2001, often shown in articles about the pilot program, is a folding-type handset, weighing 105g, with dimensions of 103H x 52W x 20D (mm) folded, and height of 180mm unfolded. The P2101V, also a folding-type is a little larger, weighing 150g, with dimensions of 104H x 56W x 35D (mm) folded. It includes a built-in digital video camera. When folded, a small display on the outside of the case called a *private window* allows users to see time, date, and information about incoming email or phone calls. An infrared wireless port can exchange data. Both handsets have jacks for carphone/mics, large colour displays, and a FOMA (IC) card slot. The handsets allow for what NTT DoCoMo calls multi-access, meaning that, while talking on the phone about how to get somewhere, both users can simultaneously download the same Internet map of the area to assist them.

The other carriers' handsets have features useful in language classes. All handsets have a similar battery life – about two to two and a half hours of use, almost double the FOMA models. One Sony handset (Sony's C413S for au) can communicate wirelessly to a Sony PC and digital video camera, and some handsets have memory card slots allowing for data (text and photo) exchange, mostly used for MP3 files. Sanyo's SCP-6000 has voice-activated dialling and Web access (*Newsweek*, July 30, 2001, p. 5). Handsets are generally quite sturdy, with two models (NTT DoCoMo R691i Geofree and au Casio's C409CA) waterproof and tested to withstand a fall from three stories high (*K-tai Oh*, 2001, pp. 34-35). It is, however, in the area of looks that Japanese designers have excelled. The displays are large, some with TFT 65,536 colour displays, the same as in top quality laptop computers. Thomas Fellger, wireless consultant at MetaDesign of Berlin, stated, "People are going to go crazy just for the [Japanese] handsets, with their big colour displays. The Europeans haven't seen anything like it" (Emerson, 2001, p. 23). The handsets are sleek, and in fashionable colours – masculine silvers, blacks, and oranges, and feminine pastels – many resembling top fashion brand cosmetics cases. This design helps account for the approximately 20 million users in their teens and twenties using NTT DoCoMo alone, as reported above (Emerson, 2001, pp. 24-25).

Three other hardware pieces need to be introduced. The first is the MM QUBE2 Office Mobile Phone Network all-in-one server, which can connect mobile phones, notebook computers, and PDAs to each other, useful hardware for an educational setting (*DoCoMo Mobile Multimedia Catalog*, 2001, p. 27). The second is the Micro Ms EM-Mode Handsfree Mic/Headphone. The microphone picks up sound from ear bones, and is part of the supporting structure of the headset and thus almost invisible, making it easy to hear callers in the classroom, and allowing the user to simultaneously look at the display screen and listen and talk at the same time (*K-tai Oh*, 2001, pp. 134-135). The third piece of hardware is the head-mounted display for the Hitachi wearable personal computer built in partnership with U.S. wearable computer maker Xybernaut. The display, which can be embedded in sunglasses, allows the user to see an image equivalent to a 13-inch monitor 60 cm away (*The Japan Times*, August 8, 2001, p. 15).

4.4.2 Present software

Besides design, a major factor contributing to the popularity of mobile phones is their capabilities, the software. A huge benefit lies in the fact that most Japanese mobile phones connect to the Internet. NTT DoCoMo, in 1999, introduced the first Internet phone service, called i-mode, allowing email, Web-surfing, and downloading of software applications (Emerson, 2001, pp. 24-25). It used html so that its software would allow customers to use the Internet easily. Most importantly, it set standards for handsets (Emerson, 2001, pp. 24-25). Therefore, the most important software is that which allows the handset to connect with the Internet. In addition, included in almost all mobile phones are a scheduling appointment book, telephone and email phone book, and a memo book, most of the basic functions of a PDA. From the Internet, mobile phone owners can download games, screen savers, and other fun applications. Some handsets include MP3 players to play music downloaded from the Internet, or stored on memory cards that can be inserted.

4.4.3 Present services

Most mobile phone service and handsets are second generation (2G) mobile phones. Many advanced areas use second-and-a-half (2.5G) generation phones. In the world now, only Japan and South Korea boast third generation (3G) mobile phones in small areas. Capabilities of different generations of mobile phones are shown in Table 1.

Table 1. Capabilities of generations of mobile phones (from Faroohar, 2001, p. 21).

Generation	2G	2.5G	3G
Status	Technology of most current digital mobile phones	Best technology now widely available	Combines mobile phone, laptop PC, and TV
Features	Phone calls Voice mail Receive simple email messages	Phone calls Fax Voice mail Send/Receive large email messages Web browsing Navigation/maps News updates	Phone calls Fax Global roaming Send/Receive large email messages High speed Web browsing Navigation/maps Videoconferencing TV streaming Electronic agenda meeting reminder
Speed	10Kbps	64-144Kbps	144Kbps-2Mbps
Download time for 3 min MP3 Song	31-41 mins	6-9 mins	11 secs-1.5 min

The services presently common in Japan are phone calls, fax, sending/receiving large email messages, Web browsing, navigation/maps for finding the way when lost or for locating a destination, and news updates. The carrier au provides voice mail. DoCoMo's i-mode now offers AOL i-service which allows more varied content and longer email messages that can be read on the handset or on computers. TU-KA and J-PHONE offer a service called Skymail which allows regular short mail. Extra services include: Coordinator Mail, which allows one person to send messages to a group; Relay Mail, which allows a person to send a message to the next person who adds to the message and sends the extended message to the next, and so on, much like a forwarded email; Hotline, which is automatic one-touch one-to-one mail; and Greeting, which allows the user to write a message and have it sent at a set time, for example on a birthday or other annual celebration. The carrier au provides chat capabilities. J-PHONE allows one to create, upload, and see a homepage on one's mobile phone. Several carriers allow the user to know in advance who is calling or mailing by using different screen or lamp colours.

One interesting non-Internet related service is prepaid mobile phones. Inexpensive prepaid-type handsets, purchased at convenience stores, have no monthly fees; instead, the user buys a prepaid card for a set price, registers the purchase with a number on the card, and can then use the phone for a set period (usually two months) for a certain amount of talk time. When the set period or allotted time is finished, the user can purchase a new card. This has possible relevance to the problem of who pays the price of mobile phones in the classroom. One prepaid-type handset, J-PE02 by Pioneer, has touchscreen input with double or triple the typical display screen size. This may alleviate the problem of displays too small to read English easily.

Problems

The Japanese mobile phone system is good, but not perfect. The main problems are cost, speed, and content. Although NTT DoCoMo is first in the world in data transmissions, it is also has the highest monthly user fee, with Sprint from the US in second place (Emerson, 2001, p. 24).

The Japanese data transfer infrastructure is very narrow. It takes between 6 and 41 minutes to download a three-minute song. With high data transfer costs, it is sometimes cheaper to buy an entire CD than download one song (Faroohar, 2001, p. 21). This should be alleviated when Yahoo!BB, Japan's first broadband service, becomes widespread and data transfers are increased, sometimes by up to 143 times the speed of analogue transfer time (*Nikkei Mobile Yahoo! Japan* advertising insert, 2001, p. 9).

The content available for download is a problem now because Japan has decided to concentrate on entertainment. When i-mode was started in 1999, NTT DoCoMo saw it mainly as a tool for businessmen, comprising stock market reports, traffic information, and online banking. The company was surprised to discover that half its users were in their teens and 20s, and the downloaded favourites were cute cartoon characters, games, and fortune-telling services (Emerson, 2001, p. 24). This situation is not changing.

Some people look to the coming broadband infrastructure and the spread of 3G mobile phone services. However, there are problems to overcome. In May 2001, NTT DoCoMo implemented a pilot 3G system called *FOMA* in Tokyo that was not very successful. A DoCoMo survey found that 47 percent of the pilot group believed the quality of FOMA connection to be poor, and 53 percent said that disrupted connections occurred more frequently than with 2G or 2.5G mobile phones. NTT DoCoMo is planning to solve this infrastructure problem by constructing more broadcast towers. One user, at first excited to be selected for the service, now rarely uses it, saying "It's next to useless," citing disrupted connections, poor battery life, and lack of content for the faster transmission speed (Yoshida, 2001, p. 14).

One large problem is a lack of standardization of handsets and services. NTT DoCoMo is the largest carrier, but handsets and services of J-PHONE and au, although popular and potentially more useful to language classes, can use only the service they were built for. One hopes that there will be a standardization resulting in inclusion of capabilities like wireless connectivity or voice mail, and services such as long email messages or chat, rather than a loss of capabilities and services.

4.4.4 Future hardware

A crystal ball may show the future HALL handset design coming about as a result of a convergence of handheld PCs, PDAs, handheld game sets, and mobile phones. From the handheld PC and the PDA will come the keyboard, large display, high processing speed, fairly large memory storage space, basic word processor, (and perhaps database and spreadsheet), and programmability. From the PDA and the handheld game set will come wireless connectivity and modularity. From the handheld game set will come interconnectivity. From PDAs and mobile phones will come basic daily functions such as scheduler, telephone number and address book, and Internet access. From the handheld PC, the PDA, and the mobile phone will come video playing capability. From the mobile phone will come phone communication, digital camera, and most importantly, size and design – a very slim, stylish device, a little bigger than the present handset, with a larger display. It has a wireless port for connectivity and interactive connectivity with other handsets, computers, printers, and digital cameras, rendering hardwired LANs and actual phone calls unnecessary for activities. It includes an internal digital video/still/Web camera, and a hands-free microphone/headphone set. The internal microphone can be used for telephoning and sound recording. It has a memory card slot for data or application exchange, and adequate memory for storage of such data and applications. It will have a long-life battery, and a processor with a speed of 206Mhz, the same as in the fastest PDA. It will have speech recognition capabilities for voice input. Finally, the mobile phone will be able to connect to a keyboard for full input.

4.4.5 Future software

The preinstalled software for future mobile phones will include a minimal word processor and database, an appointment scheduler, a Web browser, email, printer driver software, a photo/video player, and a sound recorder/player, and perhaps an English/Japanese-Japanese/English dictionary (though with fast Internet access, the present online dictionaries may be adequate). Additional software can be installed through the memory card slot, or downloaded from the Internet.

4.4.6 Future services

The service for future mobile phones will be widespread broadband service that can handle the simultaneous video and audio transmission necessary for TV phones, as well as multi-access. This will also allow the possibility of quick download of data and applications.

5. Paradigm shift

With the introduction of HALL, the CALL classroom can become a mobile, flexible, inexpensive language learning classroom. Any classroom can become a CALL classroom. This will require a rethinking of classroom activities, different from the activities of the networked desktop world of CALL and from those of the paper and book model of regular classes. This rethinking will be discussed in the next section.

6. Pedagogy

For a CALL teacher training course, the author developed five principles of computer use (Kluge, 1999) which can be adapted to handheld device use, listed below with *handheld device* substituted for *computer*:

Principle 1: Use the handheld device for activities it is good at. (*They are small, mobile, interactive, multimedia machines that use the Internet.*)

Principle 2: Do not use the handheld device for activities that a different device can do better.

Principle 3: The handheld device is just a tool to achieve the goal and is not the goal.

Principle 4: Skills taught with handheld device should be able to be done by the student without the handheld device.

Principle 5: The skill being taught is more important than learning about the handheld device.

I would like to add two more principles regarding activities:

Principle 6: Pursue activities/technologies that require little training of students.

Principle 7: Choose routine activities that are repeated often over activities which are done only once and then finished.

All principles above should be considered when creating HALL activities. How the principles are applied depends on the institution's situation and goal of using HALL. If the goal is to train students in its use, if there are no computers at the institution, or if the goal is to use HALL in all courses, then all activities are acceptable, perhaps even those just as easily done with simpler media (pencil-and-paper or tape recorder). If the goal is to complement the institution's CALL program, then only activities that can be done exclusively on the handheld device should be done.

These activities below are based on the capabilities, software, and services provided for mobile phones, the goals of language classes, and knowledge of which activities are valuable and workable in a language class. (For actual tested mobile learning projects, see Thornton and Houser in this volume.)

7. Description of activities

The activities are arranged according to skill class. They are described, then evaluated according to the number of mobile phone features used. (See Appendix A for a complete list of activities, and indication of whether it can be done now, or whether we will have to wait until a function or service becomes available.)

Let us assume we are describing an English department in a Japanese university using HALL, or what Thornton and Houser (this volume) call *m-learning*. The institution is usual in that it has a few computer labs, but most courses are taught in regular classrooms. All computers of the institution are connected to a LAN and the Internet. In this case, each classroom will have a computer with printer that can be used by mobile phones using wireless connectivity technology. In essence, we will be describing a basic HALL English language skills curriculum in Japan.

7.1 General use
First, let us look at activities done throughout the curriculum, and even outside ESL, in content classes. In all classes, students will access course syllabi and will routinely receive broadcasted administrative announcements about cancelled or rescheduled classes and homework assignments, as well as individual messages about class progress by mobile phone, either by push method via mail or by pull method via a website (Activity 1). They can receive handouts by accessing the course website, and take notes on the handout in class using a portable keyboard connected to their mobile phone (Activities 2 and 3). Students can do research for their homework using their mobile phone (Activity 4). They can do homework, then submit it via mobile phone (Activity 5). Students sometimes spend three to five hours a day commuting to and from school and could use this time productively doing homework via mobile phone.

All activities below described are evaluated according to how well they use the features of mobile phones: voice, text, and photos or video. In addition, activities are

Table 2. Evaluation of Activities 1-5.

No.	OVERALL Total=7	Voice	Text	Photo/ Video	Digital Camera	Wireless Interactivity	Mobility	Uses Internet
1	2		X					X
2	2		X					X
3	1		X					
4	2		X					X
5	2		X					X

evaluated according to how they use the built-in digital camera. Then, they are evaluated on use of mobility and interactivity through wireless technology, and whether they use the Internet. Finally, a rough overall evaluation is made, assigning one point to each category of voice, text, photo/video, digital camera, wireless interactivity, mobility, and uniqueness of technology application, for a total of seven points.

An evaluation of activities according to the features shown in Table 2 shows that none rates higher than 2, meaning that these activities do not use many features and services. Scores do not indicate how good an activity is, but how well the strengths of the mobile phone are incorporated into the activity.

7.2 All classes

All ESL classes described below, and content courses (where the emphasis is primarily on acquiring information, not skills) can use the next two activities. Quizzes or short exams would be taken on the mobile phone, with results sent immediately to the teacher's grading spreadsheet (Activity 6). Multimedia projects using audio, still shots, video, animation, and websites will use the mobile phone (Activity 7).

The evaluation of the above activities in Table 3 shows that Activity 7 fits well with the medium.

Table 3. Evaluation of Activities 6-7.

No.	OVERALL Total=7	Voice	Text	Photo/ Video	Digital Camera	Wireless Interactivity	Mobility	Uses Internet
6	2		X					X
7	5	X	X	X	X			X

7.3 The reading class

The HALL reading class would use a combination of mobile phone and, when more appropriate, a book of readings. HALL reading activities would consist of three types of purpose: reading for meaning, speed, and pleasure. The former would be cooperative learning jigsaw activities. Students, working in groups, would get the readings and worksheets via mobile phone, most likely from the course website. In one activity, all students get the same reading, but each would answer different questions on it from the worksheet. Each student would beam answers, to serve as the basis for discussion, to other members of the group via the wireless port (Activity 8). A variation on this is where each group member would get a different reading with accompanying worksheet. All readings would be on the same topic. Each person would read and answer his or her own reading and answer questions on the worksheet. Finally, each person would beam the answers to the others in the group and give a report on the reading to the group (Activity 9). Another variation is where, in groups called *home teams*, each person would either have a different reading or different worksheet questions. People from other groups with the same reading or worksheet questions would gather together in new groups, called *expert teams*, to discuss and agree on the answers. When finished, students would return to their home teams to beam the answers and discuss them (Activity 10).

When reading for speed, students would download a reading. While reading the selection, they would scroll to the next screen at a set signal or set rate. Then they would be quizzed on the reading (Activity 11). Like other quizzes (Activity 6), results would be beamed to the teacher's grading spreadsheet. An alternative would be students checking the start and finish time using the clock on their phone; dividing the total words of the reading by the time it took to read would determine reading speed in words per minute (they could use the phone's built-in calculator).

When reading for pleasure, students would receive an instalment of a story each class either via email or downloaded from the website. This could be a cliffhanger style, with the ending of each instalment leaving the hero or heroine in a life-threatening situation (Activity 12). This would not be quizzed or graded.

Other typical reading activities could be done using mobile phones. For a sample reading vocabulary activity, see Thornton and Houser (this volume). Of course, all the general activities (Activities 1-7), including quizzes by mobile phone, will be done in the reading class.

An evaluation of activities in Table 4 shows that only Activities 8, 9, and 10 use more than two features of the medium.

Table 4. Evaluation of Activities 8-12.

No	OVERALL Total=7	Voice	Text	Photo/ Video	Digital Camera	Wireless Interactivity	Mobility	Uses Internet
8	3		X			X		X
9	3		X			X		X
10	4		X			X	X	X
11	2		X					X
12	2		X					X

7.4 The speaking class

Of all language skills classes, speaking has usually used activities that require mobility and interactive exchange of data, and therefore is perhaps best suited for HALL. Two types of speaking classes will be described, the conversation class and a speech/ discussion/debate class. Both classes would use HALL as well as non-technical activities. Only the HALL activities will be discussed in this paper.

In the conversation class, students would do information gap activities. Half the class would download half the information necessary for a task in the form of a worksheet with half the information missing. The other half of the class would download the other half of the information. In pairs they would converse to fill in the entire worksheet on their mobile phone (Activity 13). Opinion or reasoning gap exercises could also be done this way.

Oral surveys (Activity 14) could be done the same way as Activity 13. The class is divided into groups, with each group selecting a topic. They create a survey on their mobile phones and create a database on the class website. Each group member asks the questions to a section of the class, recording results on a mobile phone and beaming these to a database on the class website.

One activity for describing clothing, people, or personalities is called Who Am I? (Activity 15). In this activity, one half of the class are *callers*, the other half are *receivers*. The callers sit in a group facing the wall so the receivers cannot see them talking on the phone. Each caller calls a receiver who asks questions to try to identify the caller. When the receiver identifies the caller, either both can retire from the game, or the caller can repeat the process with a different receiver. After a certain period of time, the roles change.

The next activity (Activity 16) can be used to practice giving directions. One function of present mobile phones is the ability to download maps for navigation purposes. In pairs, students download a map of a location they are interested in. One of the pair indicates the starting point and gives directions to a spot on the map. The other person guesses the destination. A modification of this activity, something akin to hide-and-seek, can be done in pairs. One of the pair hides on campus and the seeker calls the person hiding and gets directions in English on how to get to where the person is. Groups of Japanese students often do this when trying to meet up for an evening out.

Evolving roleplays (Activity 17) is similar to the usual roleplay in that each student in a pair gets the relevant information for the role, in this case downloaded from the class website. The difference is that at certain periods new information is sent to the participants. Also, decisions made by participants are sent wirelessly to the other's handset, causing changes to that person's role. The roleplay ends when both participants are satisfied with the results, or after a certain period of time. Or, the data of the first roleplay can be beamed to a new person who wants to enter, making this a never-ending activity. The same roleplay can then extend over several class periods, or even a whole semester.

The speech class will cover speeches, discussions, and debates (Activities 18-20). Random impromptu speech topics are either sent to or downloaded by students (Activity 18), who have to do a short speech on that topic. Discussion and debate (Activities 19 and 20) will be done the same way, with students doing research and putting it on the class website or using wireless technology to transfer notes to others in their group. The discussion or debate will use this information.

Of course, all the general activities (Activities 1-7), including quizzes by mobile phone, will be done in both conversation and speech classes. An evaluation of the activities shown in Table 5 shows that only Activity 17 is a good fit with the medium, and Activities 14, 15, and 16 use more than two of its features.

7.5 The listening class

If all students in a listening class are of the same level, the teacher could use a cassette tape, CD, or video player for the listening selection. However, if the HALL listening class has students at different levels, multiple listening selections can be downloaded

Table 5. Evaluation of Activities 13-20.

No.	OVERALL Total=7	Voice	Text	Photo/ Video	Digital Camera	Wireless Interactivity	Mobility	Uses Internet
13	2		X					X
14	3		X			X		X
15	3	X					X	X
16	3	X					X	X
17	5	X	X			X	X	X
18	2		X					X
19	2		X					X
20	2		X					X

and heard using headphones, allowing each student to have a different selection. This listening class uses a combination of individual and group activities using the mobile phone, and whole class listening using a tape or CD player.

The main HALL activity for this class is to practice listening by downloading the activities (audio alone or with video) and then answering questions on a mobile phone (Activity 21). Students can listen to a humorous message or story for enjoyment (Activity 22). Finally, students can go on a listening rally (Activity 23). In small groups, they call a number and listen to a recorded message giving hints indicating where to go. As they reach a place, they find posted there a different number to call for another recorded hint of where to go next.

Of course, all the general activities (Activities 1-7), including quizzes by mobile phone will be done in the listening class.

Table 6 shows that only Activities 21 and 23 use more than one feature of the medium.

Table 6. Evaluation of Activities 21-23.

No.	OVERALL Total=7	Voice	Text	Photo/ Video	Digital Camera	Wireless Interactivity	Mobility	Uses Internet
21	4	X	X	X				X
22	1	X						
23	3	X					X	X

7.6 The writing class

Students in this class can do many of the writing activities with mobile phones attached to a keyboard. Here is where the eyepiece display, which shows the equivalent of a 13 inch monitor, would really be helpful. Students can daily use the word processing function to write a journal that can be saved on a server (Activity 24). A modification of the journal can make a different activity (Activity 25), exchange journals, where students beam journal messages to a partner either wirelessly or by email. Topics for freewriting (a daily topic) or a list of topics for essays could be stored on a server or on the class website for student access (Activity 26).

For specific writing activities, mobile phones can be used. For example, descriptive writing can be done by describing photos downloaded from the server or from a memory card (Activity 27). Alternatively, students can go on a photo safari (Activity 28), taking their own photos using the internal digital camera, then returning to the classroom and writing a description of one of the photos.

An activity for process or chronological writing can be done using the mobile phone's scheduling function (Activity 29). Students write down their schedule for a day or two in their mobile phone scheduler, and write an essay based on the information in their scheduler. Another chronological writing activity is for students to download a map, choose a location, and write directions on how to get there from school (Activity 30). Finally, students can write survey questions, beam the questions to other students or email them, posting results on the class website (Activity 31).

Of course, research papers can be done using Activity 1, using mobile phones to search a server or the Web. Again, all the general activities (1-7), including quizzes by mobile phone will be done in the writing class. Table 7 shows that Activity 28 fits the medium well, and only Activities 27, 30, and 31 use more than two of its features.

Table 7. Evaluation of Activities 24-31.

No.	OVERALL Total=7	Voice	Text	Photo/ Video	Digital Camera	Wireless Interactivity	Mobility	Uses Internet
24	2		X					X
25	2		X		X			
26	2		X					X
27	3		X	X				X
28	5		X	X	X		X	X
29	1		X					
30	3		X	X				X
31	3		X			X		X

7.7 Summary

The activities above try to adhere to the principles mentioned at the beginning of the section, as explained in Table 8.

Table 8. The principles of HALL and how they have been applied.

	Principle	How Applied
Principle 1	Use the handheld device for activities it is good at.	They are small, mobile, interactive, multimedia machines that use the Internet.
Principle 2	Do not use the handheld device for activities a different device can do better.	HALL classes will not exclude other ways to do activities.
Principle 3	The handheld device is just a tool to achieve the goal and is not the goal.	Little time will be spent in teaching the hardware and software.
Principle 4	Skills taught with handheld device should be able to be done by the student without the handheld device.	Activities are as natural as possible.
Principle 5	The skill being taught is more important than learning about the handheld device.	Skills are the prime focus of the activities.
Principle 6	Pursue activities/technologies that require little training of students.	Students already know how to use mobile phones even better than teachers.
Principle 7	Choose routine activities repeated often over once-off activities done only once and then finished.	Most activities introduced here are once-off activities, but at least one routine activity is introduced for each skill.

Introducing HALL into the language classroom is a relatively inexpensive way of enabling teachers to address different student levels in skills classes without sacrificing class unity. HALL also introduces the desired mobility presently lacking in CALL classes. The above activities can be done not only in class, but also in school, town, and at home. Teachers who start using HALL will develop more activities based on better knowledge of technology, advances in technology, expansion of services, and development of new language learning software, applications, and websites. At that time, the system used in this paper can be used to evaluate new HALL activities.

8. Problems

There are several problems regarding HALL, beside the fact that some activities will have to wait for the future. The first is the cost of regular mobile phone use, alleviated if much work is done not by phone but using the wireless network and server; cost can be limited with prepaid cards. Also, the small screens will be difficult for long readings and writing. A serious problem is the very little availability for mobile phones in terms of applications and content specific to language learning. Teachers and companies will have to develop software and websites that can be downloaded onto mobile phones. Many activities above will be need a specialized server, set up with necessary data and software and maintained. Presently useful websites must be adapted for mobile phone use. Textbooks incorporating the new technology and pedagogy will have to be written to avoid the fate of TPR and CLL, methodologies marginalised due to not being brought into the orthodoxy of language teaching through the development of textbooks and course materials by the major textbook publishers. Teachers will have to become familiar and comfortable with the technology and pedagogy. Finally, the sensitive issue of teachers knowing students' phone numbers and vice versa, as well as students knowing each other's mobile phone numbers, will have to be carefully and fully addressed.

9. Conclusion

It is important to keep in mind that mobile phones are unlikely to replace computers or other technology, but rather expand the role of technology in the language classroom. If possible, a combination of computer and mobile phone is optimal. If computers are unavailable for any reason, mobile phones are a good alternative. If a mobile technology-assisted classroom is desired, mobile phones are a good solution.

To conclude, Soloway et al. (2001, p. 20) stated: "It is appropriately ironic, however, that those technologies that fit into the palm of a child's hand – the pencil, the paperback book – are really the technologies that bring about revolutions in education." Mobile phones in the language classroom in the hands of students is one technology that could bring about an education revolution. We need to be ready to embrace new technology and learn from the mistakes of early CALL practices. This paper is a step in what the Japanese call *kokoro junbi*, or "preparing our heart" for the coming innovation.

References

COMPAQ. (2001, April). *iPAQ catalog*. [Brochure]. Tokyo: Author.

Emerson, T. (2001, August 6). The next big thing? *Newsweek*, 23-25.

Eye to the future. (2001, August 8). *The Japan Times*, p. 15.

Eye to the future. (2001, August 25). *The Japan Times*, p. 8.

Faroohar, R. (2001, May 28). The other bubble. *Newsweek*, 18-23.

Handspring. (2001). *VISOR catalog*. [Brochure]. Tokyo: Author.

J-PHONE (2001, July). *J-SH07 catalog*. [Brochure]. Tokyo: Author.

J-PHONE. (2001, July). *J-PHONE catalog*, July, 2001. [Brochure]. Tokyo: Author.

J-PHONE. (2001). *J-Sky summer book*. [Brochure]. Tokyo: Author.

K-tai Oh. (2001, August), *KK Bestsellers. Vol. 2*.

KDDI. (2001, July). *au, KDDI catalog*. [Brochure]. Tokyo: Author.

Kluge, D. (1999). *Computers in English education*. Nagoya: Asano Books.

Newsweek. (2001, June 11). Design trends. *Newsweek*, 59.

Newsweek. (2001, July 9). A new threat to homework. *Newsweek*, 8.

Newsweek. (2001, July 30). Coming to a pocket near you. *Newsweek*, 5.

Nikkei Mobile. (2001, September), No. 29.

NTT DoCoMo. (2001, Summer). *DoCoMo mobile multimedia catalog Summer 2001*.

[Brochure]. Tokyo: Author.

NTT DoCoMo. (2001, July). *DoCoMo catalog*. [Brochure]. Tokyo: Author.

Palm. (2001). *Palm computing catalog*. [Brochure]. Tokyo: Author.

Sharp. (2001). *Zaurus MI-E1 catalog*. [Brochure]. Tokyo: Author.

Sharp. (2001). *Zaurus MI-EX1 catalog*. [Brochure]. Tokyo: Author.

Soloway, E., Norris, C., Blumenfeld, P., Fishman, B., Krajcik, J., & Marx, R. (2001, June). Hand-held devices are ready-at-hand. *Communications of the ACM 44, 6*, 15-20.

Sony. (2001). *au by KDDI Sony catalog*. [Brochure]. Tokyo: Author.

Sony. (2001). *Clié PEG-S500C•PEG-S300 catalog*. [Brochure]. Tokyo: Author.

TU-KA. (2001, July). *TU-KA catalog*. [Brochure]. Tokyo: Author.

Yoshida, R. (2001, August 17). DoCoMo's 3G service disappoints trial users. *The Japan Times*, p. 14.

20th anniversary: How IBM's 5150 fueled the revolution. (2001, August 9). *The Japan Times*, p. 15.

Appendix A

Table of possible activities

No.	Skill	Activity	Present/ Future	Routine/ One-Off	Description	Overall Rating (Total=7)
1	General Use	Get announce-ments	Present	Routine	Teacher or school either sends emails or posts announcements about homework, class cancellations, or rescheduling to a website.	2
2	General Use	Handouts in class	Present	Routine	Teacher either sends emails or posts handouts for the class to a website	2
3	General Use	Take notes in class	Future	Routine	Students take notes in class using a portable keyboard	1
4	General Use	Research	Future	Routine	Using server or Internet to get information for research	2
5	General Use	Do and send homework	Future	Routine	Students do and then send homework to the teacher	2
6	All	Quiz/Exam	Present/ Future?	Routine	Students take a quiz, and the results are immediately sent to the teacher's grading spreadsheet.	2
7	All	Multimedia projects	Future	One-Off	Sound, animation, photo, and video capabilities are used to make multimedia projects. This includes making and uploading websites.	5

No.	Skill	Activity	Present/ Future	Routine/ One-Off	Description	Overall Rating (Total=7)
8	Reading	Jigsaw 1	Present	One-Off	Students in groups are sent the same reading but each student answers different questions on the reading. The answers to the questions are sent to all group members. Readings and questions can be on a server or website.	3
9	Reading	Jigsaw 2	Present	One-Off	Students in groups are sent a different reading on the same topic and then answer questions on it. Answers are sent to all group members. Readings and questions can be on a server or website.	3
10	Reading	Team Jigsaw	Present	One-Off	Students in groups are sent a different reading on the same topic. All students from other groups with the same reading gather together and discuss and answer questions on it. Answers are sent to all members of the original group.	4
11	Reading	Speed Reading	Present	Routine	Students read a message while scrolling so they increase their reading speed. The text is at a variety of reading levels. Readings can be on a server or website.	2
12	Reading	Written message a day	Present	Routine	Students are sent one message a day to read for enjoyment and edification.	2
13	Speaking	Information Gap	Future	One-Off	All students get a set of information. Half the students have half the answers, the other half have the rest of the answers. They exchange information.	2
14	Speaking	Oral surveys	Present	One-Off	Questions for a survey are sent to each student to use in asking the survey questions of the other people in the school.	3

No.	Skill	Activity	Present/ Future	Routine/ One-Off	Description	Overall Rating (Total=7)
15	Speaking	Who am I? 50 questions	Present	One-Off	Half the class call the other half. The person called must identify the caller by asking questions. The caller tries to mask his/her voice.	3
16	Speaking	Where am I?	Present	One-Off	Students work in pairs. One of each pair goes somewhere on campus and calls the other, telling that person how to find the caller.	3
17	Speaking	Evolving Roleplays	Future	One-Off	Students are given information for a roleplay character. As the roleplay goes on, more information is sent. Mobile phones send the new situation to each other which changes the roleplay.	5
18	Speaking	Speech	Present	Routine	Random topics for speech are sent to each person on the list at a certain time. The student speaks on the topic received.	2
19	Speaking	Discussion	Present	One-Off	Information used for a discussion is available on a server.	2
20	Speaking	Debate	Present	One-Off	Information used for a debate is available on a server.	2
21	Listening	Listening	Present	One-Off	All listening done now could be on a server, downloaded by each student, allowing different levels of students in each class.	4
22	Listening	Recorded message a day	Future	Routine	Students are sent a voice message a day to listen to for enjoyment and edification.	1
23	Listening	Rally	Present	One-Off	In teams, students receive oral directions on how to get to the next place in the rally.	3
24	Writing	Journal	Future	Routine	Students daily write a journal that can be saved on a server.	2

No.	Skill	Activity	Present/ Future	Routine/ One-Off	Description	Overall Rating (Total=7)
25	Writing	Exchange journal	Present	Routine	Students write a journal and send it daily to a partner as mail. They also automatically send it to a server.	2
26	Writing	Topics	Present	Routine	Random topics for writing stored on a server or website are accessed for writing done on the mobile phone.	2
27	Writing	Describe photos	Future	One-Off	Students write descriptions of photos found either on servers on or a memory card obtained from the teacher.	3
28	Writing	Photo safari	Present	One-Off	Students take digital photos of various people, places, and things with the mobile phone's internal camera, and then write about the pictures either in class or at home.	5
29	Writing	Daily schedule	Present	Routine	Students write their journal using the scheduler to record what happens.	1
30	Writing	Give written directions	Present	One-Off	Students write directions on how to get to where they are, and send the directions to another student.	3
31	Writing	Written surveys	Present	One-Off	The students send written surveys to other students and send the results to others in their group with the same set of questions. The results of these surveys can be posted on a website.	3

Appendix B

Websites of products mentioned in paper

au: www.au.kddi.com

Compaq's iPAQ Pocket PC: www.compaq.co.jp/products/handhelds/pocketpc

Hasbro's Pox: www.p-o-x.com

HP's jornada 525: www.jpn.hp.com/jornada

J-PHONE: www.j-phone.com

Micro Ms EM-Mode Handsfree Mic/Headphone: www.micro-ms.com

NTT DoCoMo: www.docomo-tokai.co.jp/

Palm's m505: www.palm-japan.com

Sharp Zaurus MI-L1: www.ezaurus.com

Sony's Clié PEG-N700C: www.sony.co.jp/CLIE/

Springboard's Visor Edge: www.handspring.co.jp

TU-KA: www.tu-ka-tokai.co.jp

Index

About the authors

Dr. Charles Adamson, professor at Miyagi University, Sendai, has worked with computers and programming since 1958. He wrote the first CALL program for learning Japanese and has authored articles and given presentations on various aspects of CALL.

Yoshihiko and **Shizuko Ariizumi** are interested in action research, andragogy, and agentive psychology in addition to the development of technology-based assessment systems.

Frank Berberich teaches English and information science, and is interested in multimedia applications development, intelligent CALL systems, and educational media for developing countries. He is currently involved in establishing an NGO for educational support in Ethiopia.

Laurence M. Dryden and **Michelle Henault Morrone** have collaborated on MI- and CALL-related projects in Japanese universities since 1995.

James Duggan has recently completed a 4-year term as a research fellow at the Foreign Language Research Institute of Dokkyo University, and has been looking into teacher resources in technology, both traditional and nontraditional, most recently in the areas of Internet security and privacy.

Malcolm Field is currently working at Waseda University in Tokyo. He is particularly interested in the influences of culture on learning through ICT. At present, he is investigating whether ICT in education is promoting or reducing technological divides and fears.

Chris Houser researches human-computer interaction, designing and testing software that is easily learnt and efficiently used. His research interests include web and mobile phone interaction techniques for motivated and efficient education. Chris teaches computer animation and programming at Kinjo Gakuin University in Nagoya.

Douglas S. Jarrell has 25 years of experience teaching English in Japan. His research centres on using English as a tool for communication, hence his interest in CMC and network-based language teaching.

Kenji Kitao is a professor at Doshisha University, Kyoto. He has written numerous textbooks and books related to using the Internet for language teaching.

David Kluge started out in CALL in 1990 at Kinjo Gakuin University, Nagoya. His research interests are developing and testing a model of computer use, establishing an online Technology Assisted Language Learning and Teaching Research Institute, and researching the use of Personal Digital Assistants and mobile phones in language learning.

Michael Kruse graduated from Sussex University and is a graduate member of the British Psychological Society. He has taught in Britain, Egypt, and Japan.

Paul Lewis graduated from the University of Brighton with an MA in Media Assisted Language Teaching and Learning, and is an associate professor at Aichi Shukutoku University, Nagoya. He was co-chair and proceedings editor for the JALT CALL SIG conferences in 1997 (*Basics & Beyond*), and 1999 (*CALLing Asia*). He also edited the 1998 volume of papers entitled *Teachers, Learners, and Computers: Exploring Relationships in CALL.*

John Paul Loucky works at Seinan Women's University, and is researching into electronic dictionary use.

Kazunori Nozawa is a professor at Ritsumeikan University, teaching EFL and educational technology. He was co-founder of JALT CALL SIG and is co-editor of *CALL EJOnline.*

Kevin Ryan teaches most of his classes at Showa Women's University and the University of Tokyo with computers. He has taught in his native Chicago, Barcelona, Nanjing, and, for the last 20 years, in Tokyo. He started using hypertext in class three years before the web went world wide.

Bradley Saunders has taught in the Middle East, South-East Asia, and Europe. His research interests lie with the use of technology in the teaching of writing and vocabulary acquisition. He is pursuing a PhD from the University of Essex and teaches at Zayed University, Dubai, UAE.

Stephen Shucart is currently an associate professor of English at Akita Prefectural University in Honjo, Japan, where he runs a CALL lab with 70 iMac computers.

Monika Szirmai teaches at Hiroshima International University. She has been teaching English for more than 15 years in Hungary and Japan. Her research interests include corpus linguistics, lexicography, and CALL.

Patricia Thornton is an associate professor at Kinjo Gakuin University in Nagoya, Japan. Her recent interests include mobile technology for language learning and Web-based education in the Japanese context.

Peter Wanner has a MS in Linguistics from Georgetown University. He teaches at Kyoto Institute of Technology using CHILDES software to assist students in analysing their data.

LANGUAGE LEARNING AND LANGUAGE TECHNOLOGY
ISSN 1568-248X

1. ICT and Language Learning: A European Perspective
   Edited by Angela Chambers and Graham Davies
       2001. ISBN 90 265 1809 9 (Hardback Edition)
              ISBN 90 265 1810 2 (Paperback Edition)
2. The Changing Face of CALL: A Japanese Perspective
   Edited by Paul Lewis
       2002. ISBN 90 265 1934 6 (Hardback Edition)
              ISBN 90 265 1935 4 (Paperback Edition)